VOLUME ONE HUNDRED AND TWENTY THREE

ADVANCES IN
COMPUTERS

VOLUME ONE HUNDRED AND TWENTY THREE

ADVANCES IN
COMPUTERS

Edited by

ALI R. HURSON
*Missouri University of Science and Technology,
Rolla, MO, United States*

ACADEMIC PRESS

An imprint of Elsevier

Academic Press is an imprint of Elsevier
50 Hampshire Street, 5th Floor, Cambridge, MA 02139, United States
525 B Street, Suite 1650, San Diego, CA 92101, United States
The Boulevard, Langford Lane, Kidlington, Oxford OX5 1GB, United Kingdom
125 London Wall, London, EC2Y 5AS, United Kingdom

First edition 2021

ISBN: 978-0-12-824121-9
ISSN: 0065-2458

For information on all Academic Press publications
visit our website at https://www.elsevier.com/books-and-journals

Publisher: Zoe Kruze
Developmental Editor: Tara A. Nadera
Production Project Manager: James Selvam
Cover Designer: Matthew Limbert

Typeset by SPi Global, India

Working together
to grow libraries in
developing countries

www.elsevier.com • www.bookaid.org

Contents

Contributors

Tasnim Abar
COSIM Lab, Carthage University, Higher School of Communications of Tunis, Tunis, Tunisia

Mohamed-Slim Alouini
Computer, Electrical and Mathematical Sciences and Engineering (CEMSE) Division, King Abdullah University of Science and Technology (KAUST), Thuwal, Makkah Province, Saudi Arabia

Ahmad Alsharoa
Electrical and Computer Engineering Department, Missouri University of Science and Technology, Rolla, MO, United States

Anneliese Andrews
University of Denver, Denver, CO, United States

Magdy Bayoumi
Department of Electrical and Computer Engineering, University of Louisiana at Lafayette, Lafayette, LA, United States

Asma Ben Letaifa
MEDIATRON LAB, SUPCOM Tunisia, Carthage University, Higher School of Communications of Tunis, Tunis, Tunisia

Mahmoud Darwich
Department of Mathematical and Digital Sciences, Bloomsburg University of Pennsylvania, Bloomsburg, PA, United States

Sadok El Asmi
COSIM Lab, Carthage University, Higher School of Communications of Tunis, Tunis, Tunisia

Xiangbo Li
Amazon AWS IVS, San Diego, California, CA, United States

Mitchell Mayeda
University of Denver, Denver, CO, United States

Marjan Mernik
University of Maribor, Faculty of Electrical Engineering and Computer Science, Maribor, Slovenia

Igor Rožanc
University of Ljubljana, Faculty of Computer and Information Science, Ljubljana, Slovenia

Mohsen Amini Salehi
School of Computing and Informatics, University of Louisiana at Lafayette, Lafayette, LA, United States

Emna Zedini
Computer, Electrical and Mathematical Sciences and Engineering (CEMSE) Division, King Abdullah University of Science and Technology (KAUST), Thuwal, Makkah Province, Saudi Arabia

Preface

Traditionally, *Advances in Computers*, the oldest series to chronicle of the rapid evolution of computing, annually publishes several volumes, each one typically comprised of four to eight chapters, describing new developments in the theory and applications of computing.

The 123rd volume is an eclectic volume inspired by recent issues of interest in research and development in computer science/computer engineering, and several closely related disciplines. The volume is a collection of five chapters as follows:

Chapter 1 entitled "Downlink resource allocations of satellite-airborne-terrestrial networks integration" by Alshaora et al. studies the potential of improving data rate of ground users through integration of terrestrial, airborne, and satellite stations, and consequently establishing dynamic downlink wireless services in remote or infrastructure-less areas. In this integrated platform, however, several challenges such as the management of the resources need to be addressed. This article tackles challenges in resource management by formulating and solving optimization problem to find the best high-altitude platforms' location, access and backhaul associations, and transmit power allocation. The chapter also enumerates the advantages of the proposed scheme.

Chapter 2 articulates the crucial role of software testing techniques in detecting faults and hence reducing the risk of using it. Mayeda and Andrews in "Evaluating software testing techniques: A systematic mapping study" perform a systematic mapping by looking at and categorizing 335 studies based on several classification schemes to show the research gaps in each category. Links between research efforts and categories are also established to allow researchers to identify all of the testing technique evaluation research with interested properties. To demonstrate how researchers can use this mapping study, a small case study in evaluating the effectiveness of a black-box testing technique is presented.

Rožanc and Mernik, in Chapter 3 entitled "The screening phase in systematic reviews: Can we speed up the process?" define the notion of "Systematic Reviews" as a classification of a large number of carefully selected research reports. Naturally, selections of proper set of research reports are a delicate and crucial aspect of this method. As a common sense, such a screening process needs to be automated; a rigorous approach is described and a tool to efficiently support such an approach is presented. The effectiveness of the proposed approach is demonstrated by comparing its result against a classic manual approach.

Chapter 4 entitled "A survey on cloud-based video streaming services" by Li et al. focuses on video streaming. Flexibility and ubiquity of video streaming is due to several technologies including techniques borrowed from video compression, cloud computing, and content delivery network. This chapter is intended to survey two important aspects of video streaming: (i) Video encoding, transcoding, packaging, encryption, and delivery processes, and (ii) the way cloud computing and cloud services, as enabler of video streaming services, are employed. Finally, the chapter discusses research challenges in video streaming.

Finally, in Chapter 5 entitled "User behavior-ensemble learning based improving QoE fairness in HTTP adaptive streaming over SDN approach" Abar et al. concentrates on Quality of Experience for multimedia services. The authors argues that the quality of experience for multimedia services is possible by the interaction among system, context, and human factors. While system and context variables are predictable, the human variables are hardly predictable and hence has not been fully explored. Dynamic adaptive streaming over HTTP for a DASH framework has been developed as an efficient innovation for video streaming with the main challenge of limited server bandwidth in a multiuser platform. The chapter proposes Ensemble Learning approach based on user behavior to allocate the bandwidth collaboratively for multiusers using MPEG DASH technique to improve the visual quality of each user. Using network simulator, performance analysis in conducted showing significant gain in terms of perceived QoE fairness.

I hope that readers find this volume of interest, and useful for teaching, research, and other professional activities. I welcome feedback on the volume, as well as suggestions for topics of future volumes.

ALI R. HURSON
Missouri University of Science and Technology,
Rolla, MO, United States

CHAPTER ONE

Downlink resource allocations of satellite–airborne–terrestrial networks integration

Ahmad Alsharoa[a], Emna Zedini[b], and Mohamed-Slim Alouini[b]

[a]Electrical and Computer Engineering Department, Missouri University of Science and Technology, Rolla, MO, United States
[b]Computer, Electrical and Mathematical Sciences and Engineering (CEMSE) Division, King Abdullah University of Science and Technology (KAUST), Thuwal, Makkah Province, Saudi Arabia

Contents

Advances in Computers, Volume 123
ISSN 0065-2458
https://doi.org/10.1016/bs.adcom.2021.01.005

Abstract

This chapter studies the potential improvement in the Internet broadband of the data rate available to ground users by integrating terrestrial, airborne, and satellite stations. The goal is to establish dynamic downlink wireless services in remote or infrastructure-less areas. This integration uses satellite and high-altitude platforms (HAPs) the exosphere and stratosphere, respectively, for better altitude reuse. Hence, it offers a significant increase in scarce spectrum aggregate efficiency. However, managing resource allocation with deployment in this integrated system still faces difficulties. This chapter tackles resource management challenges by formulating and solving optimization problem to find the best HAPs' location, access and backhaul associations, and transmit power allocation. Finally, we show how our results illustrate the advantages of the proposed scheme followed by some potential future works.

1. Introduction and motivation

1.1 Background

In the past few years, the traditional terrestrial wireless networks have experienced a tremendous deluge in the number of users connected to the Internet and their demands for a variety of different services with high reliability, low latency, and high data rate [1–3]. To cater for these increasing requirements, fifth generations (5G) and beyond 5G networks are expected to provide a novel communication infrastructure that integrates new technologies and services aiming to offer ubiquitous Internet broadband global network coverage considering high achievable data rates that can reach to Tbit/s [1]. These services include Internet-of-Things, enhanced mobile broadband communications, automated driving, and massive machine type communications and will fundamentally impact societies and revolutionize the way people interact, work, and live. The combination of satellite, aerial, and terrestrial stations into an integrated single wireless network, which is named as space, air, ground (SAGIN) as shown in Fig. 1, has become of utmost importance [4]. Due to their advantages in providing high throughput, large coverage and strong resilience, SAGIN are appealing for a wide application ranges such as observation and mapping of the earth, Intelligent communication systems (such as intelligent transportation systems (ITS)), disaster recovery, and military mission among others [4].

The integration of the three network segments (i.e., space, air, and ground) aiming to achieve seamless and reliable Internet connectivity services with high data rate at any place in the earth, especially in rural, ocean, and mountain areas where the use of optical fiber is very costly. Indeed,

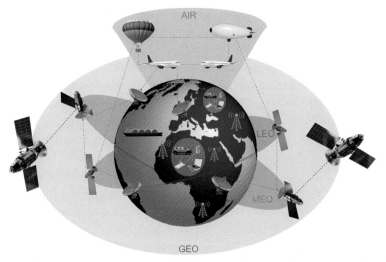

Fig. 1 Global connectivity system architecture of space, air, ground integrated network (SAGIN).

according to recent statistics, only around 55% of the world's population has access to the Internet, where most of the unconnected population is typically located in world's poorest regions, in which the infrastructure is scarce [5]. From this perspective, connectivity to these nonserved and under-served areas of the world would be ideally reached by means of broadband multi-beam satellite systems that belong to the space network.

In general, the satellite stations are categorized into three categories depending on the type of the orbit (i.e., altitude) as: (a) geosynchronous orbit (GEO), (b) medium-earth orbit (MEO), and (c) low-earth orbit (LEO). Among all satellite types, the GEO satellites can be used to provide the largest communication coverage, while the LEO satellites can be used to provide the smallest communication coverage [4]. Depending on the users' mobility, the current space communications can offer two services. The first service is a fixed space service (FSS), that is mainly focus on providing services to users in relatively-fixed locations [6]. This may consists of satellite to satellite link or satellite to fixed ground gate ways link. The other service is a mobile space service (MSS), which allows dynamic and mobile users to have voice and data communications through satellite stations [7]. In recent years, several space constellations/projects have been proposed aiming to achieve global Internet broadband access. All the proposed constellations/ projects are based on the conventional Radio Frequency (RF) solutions that operates at the Ku-band (12–18 GHz), the Ka-band (27–40 Ghz), and the

Q/V band (40–50 GHz). For example, GEO satellite station, such as KA-SAT owned by Eutelsat aims to provide high throughput around 70 Gbit/s, Viasat 1, Viasat 2, and Viasat 3 owned by ViaSat Inc. and Telesat Canada are expected to provide 140 Gbit/s, 350 Gbit/s, 1 Tbit/s, respectively [8]. Another example, is using large number of LEO satellite to provide high throughput, such as the Starlink project that supported by SpaceX and using 12000 LEO satellites [9], Oneweb project using 900 LEO satellites [10], and Telesat project using around 400 satellites [11].

However, both services are constrained mainly because of high pathloss attenuation. Apart from that, the location of the satellite stations in different orbital heights can cause significant delays when providing real-time services. However, several challenges need to be overcome first such as: (a) providing latency sensitive and prohibitive feedback load for large number of users, (b) receiver's high cost, and (c) accessing to all aforementioned developed international projects, dose not guarantee access-everywhere. This is because accessing requires landing frequency rights that might not be authorized in some countries [12].

Terrestrial-satellite integration was one of the proposed ideas to increase the network capacity and provide reliable communications. In such cooperative platform, the terrestrial base stations feed the satellite stations via high capacity link (called feeder link). Then, the satellite stations communicate the signal to ground users using multiple beams (called user link) [8]. Various integration models have been proposed in the literature to improve the feeder and user links that includes proposing multiple beams spot associated with different label switching and bandwidth control access protocol [13,14]. In such integration, the satellite stations can communicate with users in the ground using the help of the terrestrial stations. Working as relays, the terrestrial stations can broadcast/amplify the communication signals between satellite to users or vice versa. In [14], using multiple amplify-and-forward antennas, the integrated system named as hybrid satellite–terrestrial relay networks (HSTRNs) with multiple interferers at both the terrestrial relay and destination has been presented. Because each associated beam area requires different transmit power levels depending on the weather condition, dynamic steering capability, and dynamic transmit power allocations are also required to be considered. Dynamic spectrum allocation schemes have been proposed to account for hot zones, signal diversity, user priority, service types, and adverse weather conditions. To reach an optimal spectrum utilization, such schemes usually allocate more system resources to beams with higher traffic demand or bad channel conditions [15–17]. For example, Li et al. [17] introduced the hunger marketing spectrum optimization strategy to address the unpredictable user demands in multibeam satellite systems.

Although satellite–terrestrial integrated networks can provide largest coverage, it adds another limitation especially in remote or infrastructure-less areas (i.e., in the case of absent of the terrestrial networks) and during the period of congestion (in the case of fully loaded terrestrial networks). To overcome this limitation, integrating the satellite station with higher altitude stations in addition to terrestrial networks can be a promising solution for achieving global connectivity. Table 1 summarizes the differences between the three network segments [4,18]. High-altitude platforms (HAPs) such as aircrafts, airships, and balloons, which operate in the air network, can be used instead to provide a wide coverage area with lower transmission delay [19,20]. Situated well above the clouds in the stratosphere at an altitude of 17–22 km above the earth's surface, HAPs can combine the good features of satellite and terrestrial stations [21–23]. In this case, the HAPs can use hybrid source of energy: fuel and solar sources, where HAPs can potentially be deployed without/with minimal energy consumption. As shown in the literature, HAPs can be implemented as aerostatic platforms (e.g., balloons and airships) where a lighter-than-air (e.g., helium) is used to generate buoyancy and gasoline engines or solar energy is used to propel the HAPs. HAPs can also be implemented as long endurance unmanned aerial vehicle with self-operated powers [24,25]. The HAPs' altitude are chosen carefully to reduce the needed energy consumption to maintain the position stability of HAPs [21], where such altitude range is characterized by low turbulence and low wind.

Compared with the terrestrial stations, HAPs have the following features [19,21,26]:

- Higher communications coverage: The communications coverage area of HAPs has a radius around 30 km. While the communication coverage area of the terrestrial stations around 1 km. Therefore, few HAPs are sufficient to cover states or small countries [20].
- Dynamic deployment: HAPs have the ability of flying to certain areas (including infrastructure-less areas) to start the communications and provide on-demand services.
- Lower consumed energy: several studied showed that HAPs can be self-powered by equipping them with solar panels that collecting the son renewable energy during the daytime [24,25].

Compared with the satellite stations, HAPs have the following feature [19]

Table 1 A comparison of space, air, and ground networks [4, 18].

Segment	Objects	High above earth	One way delay	Data rates	Advantages	Disadvantages
Space	GEO	35,786 km	About 270 ms	Least performance	Large coverage, broadcast/multicast	Long propagation latency, limited capacity, high mobility
	MEO	2000–35,786 km	About 110 ms	Up to 1.2 Gbps		
	LEO	160–2000 km	Less than 40 ms	Up to 3.75 Gbps		
Air	Airship Balloon UAV	17 – 30km (HAP), Less than 10km (LAP)	Medium	High data rates	Wide coverage, low cost, flexible deployment	Less capacity, unstable link, high mobility
Ground	Cellular Ad Hoc WiMAX WLAN	N.A.	Lowest	High data rates	Rich resources, high throughput	Limited coverage, vulnerable to disaster

- Quicker deployment: deploying the HAPs is much faster that the satellite stations. Thus, HAPs accommodate unexpected traffic and temporal demand quickly, where one HAP is enough to start the Internet service. In addition, HAPs can be used in emergency services because their ability to fly to the desired area in short time and restart the communications [27].
- Lower cost: the deployment cost of HAPs is much lower than satellite stations.
- Lower propagation: HAPs have a lower propagation delays due to lower pathloss with users compared to the satellite stations [26].

Additionally, compared with the low altitude platforms (LAPs), HAPs have the following features [24,25]:

- Higher communications coverage: HAPs provide much broader communication coverage than LAPs.
- Period of operation: because that the LAPs are battery operated, the operated times of HAPs are much more than the LAPs. Additionally, the LAPs need to be charged frequently and thus they need to build charging stations or find another recharging strategy. However, with a careful trajectory, HAPs can operate for months without the need for recharging, such as Odysseus project [28], Astigan project [29], Zephyr project [30], Avinc project, [31], and Loon project [32].
- Wider in applications: HAPs can be used not only for Internet broadband, but also for other service such as observations of the earth, remote sensing military applications, real-time navigation, meteorological measurement, real-time traffic monitoring.

Several projects on HAP networks have been successfully launched worldwide including Helios in the USA, CAPANINA, HAPCOS, HeliNet and SatNEX in Europe, and SkyNet in Asia [22]. Moreover, HAPs can be used in many applications such as remote sensing, surveillance, earth observation, meteorological measurements, traffic monitoring and control, and emergency communication services [22].

1.2 Related work

Several research works on the deployment of HAP networks have been presented in the literature [24,25,33]. Taking into account the HAPs' coverage as well as interference model, the authors in [33] proposed a self-organized approaches aiming to provide high data rate of users. In this work, each HAP is modeled as self-organized and rational player. Moreover, a trajectory optimization technique for solar-powered HAPs was proposed in [24]. The authors introduced a real-time heuristic approach based on earth low complexity greedy algorithm that computes the optimal trajectory, while the energy consumption is minimized and bounded by the energy storage constraint. Furthermore, they extended their results to find the optimal state and input trajectories that maximize the energy storage of HAPs [25]. Enhancing the total network throughput has attracted great research attention in HAP-based communication systems. The authors in [34] proposed enhancing the network throughput using orthogonal frequency division multiple access (OFDMA) technique in downlink combined with multicasting. More specifically, the work formulated and solve an optimization problem that optimizes the resource allocation such as transmit powers, time slots, and subchannels aiming to maximize the number of connected users that receive the multicast request in a given OFDMA frame. In [35], the authors demonstrated a significant improvement in the capacity of HAP networks operating in the millimeter wave bands. By exploiting the directionality of user antennas, it has been shown that HAPs are allowed to share the same carrier frequency, and therefore enhance the spectral efficiency. A farther improvement can be done by using multiple-input multiple-output (MIMO) communications system with large scale antenna arrays [36]. The authors proposed a max-min beamforming algorithm to limit the cross-interference between users by maximizing the signal to interference noise ratio (SINR). Interestingly, the new proposed algorithm has been shown to outperform the traditional analog beamforming methods, being suitable for power consumption and complexity constrained HAP-based systems, as well.

However, these research works have not considered the backhaul links for the HAPs and only few studies have focused on the full integration of satellite, air, and ground networks [37,38]. One research work in [37] focuses in integrating the LEO satellite with HAPs equipped with solar panels for downlink communication. The HAP proposed to work as a relay between the LEO satellite and users in the ground aiming to minimize the total consumed energy while satisfying the user' throughput requirement. Another work in [38], introduced the concept of bidirectional communications using virtual model to chain the reconfiguration framework of the network.

1.3 Contributions

Satellite stations and HAPs can play a critical role in achieving the global connectivity, mainly in the absent of terrestrial network (e.g., remote areas) or when the terrestrial network is overloaded. This chapter study the improvement of the global connectivity by managing the downlink resources for access and backhaul links in the integrated space, air, and terrestrial networks. More specifically, the goal is to enhance the downlink data rate of the users, while taking into consideration the resource limitations in both access and backhaul links. The main contributions of this chapter can be summarized as follows:

- Proposing a complete three tiers global connectivity framework consisting of satellite, airborne, and terrestrial networks aiming to improve the network coverage.
- Formulating an optimization problem that maximizes the users' downlink data rate that managing the resource allocations in access and backhaul links while considering network's resource constraints such as: the transmit power budgets constraint, association constraints, and backhaul limitations.
- Formulating two different optimization problems noted as short-term stage and long-term stage optimization problems, based on how often we optimize the decision variables. In the short-term stage optimization problem, given fixed HAPs' location, we optimize the (i) transmit power allocation, and (ii) access and backhaul associations. In the long-term stage optimization problem, we solve for the HAPs' location depending on the average distributions of the users. More precisely, the short-term stage optimization variables proposed to be tuned frequently, whereas the long-term stage variables proposed to be tuned in long time periods. The details are given below:
  - Short-term optimization problem: Due to the nonconvexity of the formulated optimization problem, two solutions will be proposed. The first solution starts by jointly optimizes the access and backhaul associations assuming fixed transmit power. Given these associations, we then find the optimal transmit power levels using Taylor expansion approximations. While, the second solution proposes a low complexity heuristic approach based on frequency partitioning (FP) technique to jointly optimize the access and backhaul associations along with transmit power levels.
  - Long-term optimization problem: an efficient and low complexity algorithm based on shrink-and-realign (SR) technique is proposed to find the best locations of HAPs. This solution is repeated every long time frame.
- Considering different fairness functions between users.
- Analyzing the performance of the proposed solutions under various system parameters.

The remainder of this chapter is organized as follows. The system model that includes the assumptions, channel models, associations, and rate functions is outlined in Section 2. The problem formulation, which includes short-term and long-term optimization problems, is given in Section 3. The proposed solutions are presented in Section 4 including the short-term and long-term stages. Selected numerical results are provided in Section 5. Finally, Section 6 gives the conclusions and future directions.

2. System model

Let us define the following sets in the system model:

leftmargin=* Stations set: includes the information about satellite, HAPs, and terrestrial stations. Containing: the Identification number (ID) of each station, the three-dimensional space (3D) location, backhaul rate, and, if applicable (e.g., HAPs), the instantaneous battery level.

leftmargin=* Ground users set: contains the information about the users. Containing: the ID of each user, two-dimensional space (2D) location, and Quality-of-Service (QoS) requirements.

leftmargin=* Gateways set: contains the information about the gateways. Containing: the ID of each gateway, and 2D location.

In this mode, an integrated networks is considered to support ground users with downlink communications. We consider a three-tier networks: (i) space-tier consists of one LEO satellite station, (ii) air-tier consists of L HAPs, and (iii) ground-tier with M terrestrial stations as shown in Fig. 2. In addition, we consider W feeder station gateways. The gateways are used as backhaul stations for satellite and HAPs. The goal of this integrated model is to provide downlink Internet connectivity to U users distributed in a certain geographical area. Table 2, summarizes the main notations used in this chapter.

2.1 Assumptions

In this book chapter, we adopt the orthogonal frequency approach, where we divide the available spectrum bandwidth into resource blocks. For simplicity, let S denoted as the tier label, where $S = 0$ for the space-tier, $S = L$ for the air-tier, and $S = M$ for the ground-tier. In this chapter, we refer to s as the station label, either satellite station, HAP station, or terrestrial station. Each resource block has a bandwidth denoted as B^S Hz [39].

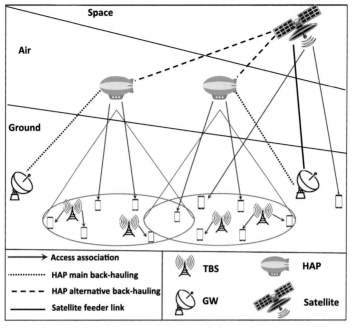

Space

Air

Ground

> Access association

.... HAP main back-hauling

– – – – HAP alternative back-hauling

————— Satellite feeder link

TBS

HAP

GW

Satellite

Fig. 2 Space, air, ground architecture system model.

The interference between different tiers are ignored because the bandwidth is divided sparsely. We denote N^S as the number of available resource blocks at tier S. In addition, we make the following practical assumptions: (i) At maximum, each user can be connected to one station using one resource block. (ii) The downlink intracell interference between users connected to the same station is ignored, where each station is able to employ orthogonal frequency approach to its connected users. (iii) We consider the intercell interference in the ground-tier, where the users connected to nearby terrestrial stations can suffer from interference if they use same resource blocks. (iv) Because of the large distance between the station in the space and air-tiers, intercell interference is ignored as satellite stations and HAPs stations can be equipped with different antennas with beams steering capabilities. Therefore, the stations in the space and air-tiers can manage their resources using different frequency sets (FSs) [35]. Note that the antenna array of HAPs is arranged to produce a regular hexagonal structure, where multiple antenna beams payload at each HAP serve multiple ground cells as illustrated in Fig. 3.

Table 2 Notations list.

Notation	Description
U	Number of users
M	Number of terrestrial stations
L	Number of HAPs stations
W	Number of gateways stations
N^S	Number resource block available resource blocks at tier S
$\mathbf{J}_u^U$	User 2D coordinates
$\mathbf{J}_s^S$	Station s 3D coordinates at tier S
$h_{su,n}^S$	Access channel gain
g_{xy}	Backhaul channel gain
A_{su}^S	Attenuation gain due to environment effects at tier S
$\Omega_0, \Omega_1, \Omega_2$	Rician shadowing parameters
κ_L	Rician fading parameter
$\epsilon_{su,n}^S$	Access association between station s and user u over resource block n
δ_{wl}	Backhaul association between gateway w and HAP l
$P_{su,n}^S$	Access transmit power of station s over resource block n
P_0	Backhaul transmit power
B^S	Access bandwidth of at tier S
B_0	Backhaul bandwidth
$\mathcal{N}_0$	Noise power value
$R_{su,n}^S$	Access data rate from station s to user u over resource block n
$\overline{R}_{ws}$	Backhaul data rate from gateway w to station s
T	Maximum shrink and realign algorithm iterations

2.2 Channel model

We consider a 3D coordinate system where the coordinate of station s and user u are given, respectively, as[a] $\mathbf{J}_s^S = [x_s^S, y_s^S, z_s^S]'$, $\mathbf{J}_u^U = [x_u^U, y_u^U, 0]'$.

---

[a] [.]' symbols denotes the transpose operator.

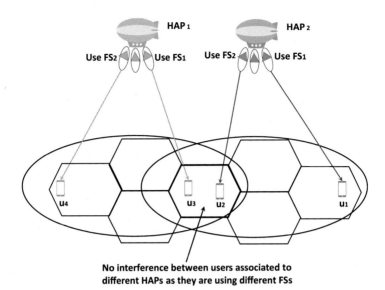

Fig. 3 Locations of the HAPs.

2.2.1 Access channel model

Considering the effects of pathloss, fading, and shadowing, the channel gain model between station s and user u can be given as:

$$h^S_{su,n} = \left(\frac{C}{4\pi d^S_{su} f^S_c}\right)^2 A^S_{su} F^S_{su,n}, \tag{1}$$

where C and f^S_c are the speed of light constant and carrier frequency for tier S, respectively. d^S_{su} is the Euclidean distance between station s in tier S and user u. In (1), the first term $\left(C/4\pi d^S_{su} f^S_c\right)^2$ is the propagation and pathloss effects where the second term A^S_{su} captures the attenuation gain due to environment effects and can be modeled as follows [34,40]:

$$A^S_{su} = \begin{cases} A^M_{mu} = 1, & \text{for ground} - \text{tier}, \\ A^L_{lu} = 10^{\frac{3d^L_{lu}\chi}{10z^L_l}}, & \text{for air} - \text{tier}, \\ A^0_{0u} = 10^{\frac{3d_{0u}\chi}{10z_0}}, & \text{for space} - \text{tier}, \end{cases} \tag{2}$$

In (2), the parameter χ denotes the attenuation factor due to environment effects such as cloud, rain and wind. The last factor in (1), $F^S_{su,n}$, corresponds to fading distribution between station s in tier S and user u over resource block n and given as follows [34,41]:

$$
F_{su,n}^S = \begin{cases} F_{mu,n}^M : \text{Rayleigh fading with a parameter } a, \text{ s.t } E\{|a|^2\} = 1 & , \text{for ground} - \text{tier,} \\ F_{lu,n}^L : \text{Rician small scale with a factor } \kappa_L & , \text{for air} - \text{tier,} \\ F_{0u,n}^0 : \text{Shadowed Rician fading with factors } \Omega_0, \Omega_1, \Omega_2 & , \text{for space} - \text{tier,} \end{cases}
$$

(3)

where the Rayleigh parameter a is selected such that $E\{|a|^2\} = 1$. For the shadowed Rician fading, Ω_0, and $2\Omega_1$ are the average power of the line-of-sight (LoS) component and of the multipath component, respectively. Ω_2 is a Nakagami parameter ranging from 0 to ∞ that shows the intensity of the fading [42]. Without loss in generality, fast fading is assumed to be approximately constant over the subcarriers of a given resource block, and independent identically distributed over resource blocks.

2.2.2 Backhaul channel model

We assume that the backhaul channel gain between x and y nodes, i.e., between gateway ($x = w$) HAP l ($y = l$) for the main backhaul link, and between satellite station ($x = 0$) and HAP l ($y = l$) for the alternative backhaul link, can be given as [34,41,43,44]

$$
g_{xy} = \left(\frac{C}{4\pi d_{xy} f_c}\right)^2 A_{xy} F_{xy},
$$

(4)

where F_{xy}, corresponds to Rician fading distribution between x and user y with Rician factor equal to κ_{xy}. The choice of κ_{xy} will be depends on the main or alternative backhaul links.

2.3 Access association

Let us introduce a new binary variable $\epsilon_{su,n}^S$, where it equals to 1 if station s is associated with user u over the resource block n and 0 otherwise, and given as follows:

$$
\epsilon_{su,n}^S = \begin{cases} 1, & \text{if station } s \text{ is associated with user } u \text{ over resource block } n. \\ 0, & \text{otherwise.} \end{cases}
$$

(5)

we assume that each ground user can be associated with only one station (i.e., either a terrestrial station, HAP station, or satellite station) using only one resource block at a certain time. Therefore, the following constraint should be respected:

$$\sum_{S \in \{M, L, 0\}} \sum_{s=1}^{S} \sum_{n=1}^{N^S} \epsilon_{su,n}^{S} \leq 1, \quad \forall u. \tag{6}$$

On the other hand, each station can be associated with multiple users. Therefore, the following constraint should be respected:

$$\sum_{u=1}^{U} \epsilon_{su,n}^{S} \leq 1, \quad \forall s, \forall S, \forall n = 1, \ldots, N^S. \tag{7}$$

2.4 Backhaul association

It is assumed that HAP l can be associated with either gateway w in the main backhaul link or with the satellite station in the alternative backhaul link based on the users distributions and the quality of the backhaul links as shown in Fig. 2. Therefore, we introduce another binary variable δ_{wl} for the backhaul association where δ_{wl} ($w = 0$ for satellite station and $w = 1, \ldots, W$ for gateways) is equal to 1 if HAP l is associated with station w and 0 otherwise, and given as follows

$$\delta_{wl} = \begin{cases} 1, & \text{HAP } l \text{ is associated with station } w. \\ 0, & \text{otherwise.} \end{cases} \tag{8}$$

Without loss of generality, we assume that each HAP should be strictly associated with one station (either one gateway or satellite station). Therefore, the following equality condition should be respected:

$$\sum_{w=0}^{W} \delta_{wl} = 1, \quad \forall l. \tag{9}$$

Finally, we denote by $\overline{L}_w$ the maximum number of HAPs that can be associated with node w in the backhaul, such that:

$$\sum_{l=1}^{L} \delta_{wl} \leq \overline{L}_w \quad \forall w. \tag{10}$$

2.5 Downlink data rates

2.5.1 Access data rates

Because the total spectrum is shared sparsely between tiers, then the achievable access data rate of user u served from a station S over resource block n can be given as:

$$R_{su,n}^S = B^S \epsilon_{su,n}^S \log_2 \left(1 + \frac{P_{su,n}^S h_{su,n}^S}{\mathcal{I}_{su}^S + \mathcal{N}_0 B^S} \right), \tag{11}$$

where $P_{su,n}^S$ is the station's transmitted power allocated to resource block n, $\mathcal{N}_0$ is the noise power, and $\mathcal{I}_u^S$ is the intercell interference at the user u caused by closest station and expressed as follows:

$$\mathcal{I}_{su}^S = \begin{cases} \displaystyle\sum_{\substack{\tilde{m}=1 \\ \tilde{m} \neq m}}^{M} \left(\sum_{\substack{\tilde{u}=1 \\ \tilde{u} \neq u}}^{U} \epsilon_{\tilde{m}\tilde{u},n}^M P_{\tilde{m}\tilde{u},n}^M \right) h_{\tilde{m}u,n}^M & \text{, for ground-tier,} \\ 0 & \text{, for air-tier,} \\ 0 & \text{, for space-tier,} \end{cases} \tag{12}$$

where $\epsilon_{\tilde{m}\tilde{u},n}^M$ is representing the exclusivity of the terrestrial station m and resource block allocation: $\epsilon_{\tilde{m}\tilde{u},n}^M = 1$, if resource block n of nearesource blocky station is allocated to another user $\tilde{u}$ from terrestrial station m, and $\epsilon_{su,n}^S = 0$, otherwise. In (12), it can be noticed that the intercell interference can only affect the users associated with terrestrial stations, because the same resource block might be allocated to the more than one user in different terrestrial stations coverage areas. On the other hand the intercell interference is equal to 0 for both air and space-tier because we assume that HAPs and satellite stations are equipped with different antennas and different beams that allow managing the resources in an efficient way by using different FSs for different ground cells [35].

2.5.2 Backhaul data rates

The backhaul data rate from station w to HAP l can be expressed as:

$$\overline{R}_{wl} = \delta_{wl} B_0 \log_2 \left(1 + \frac{P_0 g_{wl}}{\mathcal{N}_0 B_0} \right), \tag{13}$$

where B_0 and P_0 are the backhaul bandwidth and transmit power at station w, respectively.

3. Problem formulation

In this book chapter, the aim is to maximize the end–to–end users' downlink throughput by managing the available resources while taking into account the network limitations such as the maximum allowable power of the stations and users, available bandwidth, and HAPs' placement.

3.1 Resource management

The aim of resource management is optimize the resources to achieve high downlink data rate using different utility metrics. Based on the required fairness lever, the utility metric will be selected. Some examples such as (i) maximizing the sum utility, i.e., maximizing the sum data rate of all users regardless of their fairness (this may results some users enjoying high Internet broadband, while others suffering from low data rate) and (ii) maximizing the minimum utility, i.e., maximizing the minimum user throughput. This will add more fairness compare to maximizing the sum utility. In this book chapter, several parameters can be optimized to achieve this goal

- *Transmit power and bandwidth allocations*: How to optimize the transmit power allocation at the stations and users? In addition, given a certain available bandwidth, how to allocate this bandwidth among stations?
- *Associations*: Two associations are considered: (a) access association: the association between stations and users, (b) backhaul association: the association between stations and gateways.
- *HAPs placement*: How to find the best HAPs' locations taking into consideration the backhaul link quality.

The optimization problem will optimized these parameters to achieve the best utility. For example, the optimization solution determines weather be more efficient for a certain users to be connected to satellite station or HAP station or terrestrial station. For remote areas or heavy loaded areas where the terrestrial stations' would not able to support all users, the optimizer may decide to use the HAPs and/or satellite station when needed to support the ground users. This depends on many factors such as the users' distribution, number of the users to be served, and the existence/capacity of the terrestrial stations. For instance, if the users are located in remote areas (i.e., outside the coverage of terrestrial stations), then the satellite station and HAPs will try to accommodate all users in this area. In counterpart, in the case of highly loaded networks, the optimizer is forced to use the network full capacity including the HAPs and satellite station in to meet the users' demand.

3.2 Optimization problem

We now formulate our optimization problem as follows:

$$\underset{\substack{\epsilon^S_{su,n}, \delta_{wl} \\ P^S_{su,n}, \mathbf{J}^L_l}}{\text{maximize}} \quad \mathcal{U}(R_u) \tag{14}$$

subject to :

(C1 : peak power constraint :)

$$0 \leq \sum_{n=1}^{N^S} \sum_{u=1}^{U} \epsilon^S_{su,n} P^S_{su,n} \leq \overline{P}_S, \quad \forall s, S \in \{M, L, 0\} \tag{15}$$

(C2 : backhaul rate constraint :)

$$\sum_{n=1}^{N^L} \sum_{u=1}^{U} R^L_{lu,n} \leq \sum_{w=0}^{W} \overline{R}_{wl}, \quad \forall l, \tag{16}$$

(C3 : access associations constraint :)

$$\sum_{S \in \{M, L, 0\}} \sum_{s=1}^{S} \sum_{n=1}^{N^S} \epsilon^S_{su,n} \leq 1, \quad \forall u, \tag{17}$$

(C4 : access interference constraint :)

$$\sum_{u=1}^{U} \epsilon^S_{su,n} \leq 1, \quad \forall s, \forall S, \forall n = 1, ..., N^S, \tag{18}$$

(C5, C6 : backhaul associations constraints :)

$$\sum_{w=0}^{W} \delta_{wl} = 1, \quad \forall l, \tag{19}$$

$$\sum_{l=0}^{L} \delta_{wl} \leq \overline{L}_w \quad \forall w, \tag{20}$$

where $\mathcal{U}(R_u)$ denotes the data rate utility of all users. Constraint (15) represents the peak power constraints at station s in tier S. Constraint (16) represents the HAPs' backhaul data rate. Constraint (17) is to ensure that each user can be associated with one station at most using only one resource block. Constraint (18) is to ensure there is no intracell interference between users associated to the same station. Finally, constraints (19) and (20) represent the HAPs' backhaul association conditions.

3.3 Utility selection

Two different utilities are considered. The utility metrics in the optimization problem is given in (14)–(20).

3.3.1 Max-sum utility (MSU)

The utility of this metric is equivalent to the sum of users' access data rate, i.e., $\mathcal{U}(R_u) = \sum_{u=1}^{U} R_u$, as it promotes users with favorable channel gains and interference by allocating the best resources to them. On the other hand, using this utility will deprive users with bad channel gains and interference from the resources and thus they will have very low data rates [45].

3.3.2 Max-min utility (MMU)

Due to the unfairness of MSU resource allocation, the need for more fair utility metrics arises. Max-min utility (MMU) is attempting to maximize the minimum throughput of all users such as: $\mathcal{U}(R_u) = \min_{u}(R_u)$. Note that, by increasing the priority of users having lower rates, MMU leads to more fairness in the network [46]. In order to simplify the problem for this approach, we introduce a new decision variable $R_{\min} = \min_{u}(R_u)$, where $R_{\min}$ is the minimum rate among all users. Therefore, our optimization problem can be reformulated as (Eqs. 15–20):

$$\underset{\epsilon_{su,n}^{S}, \delta_{wl}}{\text{maximize}} \quad R_{\min} \qquad (21)$$
$$P_{su,n}^{S}, \mathbf{J}_{l}^{L}, R_{\min}$$

$$\text{subject to :}$$
$$R_u \geq R_{\min}, \quad \forall u, \qquad (22)$$

▷ 4. Proposed solution

The formulated optimization problem given in (14)–(20) is considered as a mixed integer nonlinear programming (MINLP). Therefore, solving it is considered as not easy task. To simplify the formulated optimization, we propose to find the solution by dividing it into two stages, namely short-term stage and long-term stage. In short-term stage, we propose to optimize the access/backhaul associations and stations' transmit power allocation. Two solutions are discussed in this stage: a near optimal solution and a low complexity solution. While in the long-term stage, we provide an efficient solution that finds the HAPs' best locations.

4.1 Short-term stage

In the short-term stage and considering fix HAPs' locations, we firstly solve for the access associations (i.e., $\epsilon_{su,n}^{S}$) and backhaul associations (i.e., δ_{wl}) using uninform power distributions, i.e., $P_{su,n}^{S} = \overline{P}_{S}/N^{S}$. We then optimize the stations' transmit power given these associations. In the sequel, we propose two solutions for this stage. In the first solution, we propose a near optimal solution based on Taylor expansion approximation. While in the second solution, a low complexity heuristic approach using frequency partitioning (FP) technique is proposed. The FP technique allow us to cancel the inter-cell interference among terrestrial stations and, thus, solve the problem quickly.

4.1.1 Near optimal solution

In fact, by fixing the transmit powers $P_{su,n}^{S}$, the optimization problem becomes linear binary optimization problem in terms of $\epsilon_{su,n}^{S}$ and δ_{wl}, except (14) which is nonlinear with respect to $\epsilon_{su,n}^{S}$ due to the existence of the $\mathcal{I}_{mu}^{M}$ term in (12). On the other hand, by fixing $\epsilon_{su,n}^{S}$ and δ_{wl}, the optimization problem is convex problem in respect to $P_{su,n}^{S}$ except constraints (14) and (16). Therefore, in order to solve the problem iteratively (i.e., solving the associations and then transmit powers), the goal becomes to linearize (14) with respects to $\epsilon_{su,n}^{S}$ using fix values of $P_{su,n}^{S}$. Then, approximate (14) and (16) to convex function with respects to $P_{su,n}^{S}$ and use the previous optimized value of $\epsilon_{su,n}^{S}$ and δ_{wl}.

Note that the term $\mathcal{I}_{mu}^{M}$ in (12) can be rewritten as $\sum_{\substack{\tilde{m}=1 \\ \tilde{m} \neq m}}^{M} \left(\sum_{\substack{\tilde{u}=1 \\ \tilde{u} \neq u}}^{U} P_{\tilde{m}\tilde{u},n}^{M} \right) h_{\tilde{m}u,n}^{M}$ by adding the following constraint to the optimization problem:

$$0 \leq P_{\tilde{m}\tilde{u},n}^{M} \leq \epsilon_{\tilde{m}\tilde{u},n}^{M} \overline{P}_{M}, \quad \forall \tilde{m}, \forall \tilde{u}. \tag{23}$$

Now, the optimization problem becomes a linear optimization problem with respect to $\epsilon^S_{su,n}$ and δ_{wl} and thus, it can be solved using on-the-shelf softwares such as Gurobi/CVX interface [47].

On the other hand, by using the previous optimized values of association variables (i.e., $\epsilon^S_{su,n}$ and δ_{wl}), the optimization problem in (14)–(20) can be rewritten as:

$$\underset{P^S_{su,n}}{\text{maximize}} \quad \mathcal{U}(R_u) \tag{24}$$

subject to :

$$0 \leq \sum_{n=1}^{N^S} \sum_{u=1}^{U} \epsilon^S_{su,n} P^S_{su,n} \leq \overline{P}_S, \quad \forall s,\, S \in \{M, L, 0\} \tag{25}$$

$$\sum_{n=1}^{N^L} \sum_{u=1}^{U} R^L_{lu,n} \leq \sum_{w=0}^{W} \overline{R}_{wl}, \quad \forall l. \tag{26}$$

Note that the optimization problem (24)–(26) is now convex optimization problem in terms of $P^S_{su,n}$ except constraints (24) and (26). Therefore, the goal is to convert these constraints into convex ones in terms of $P^S_{su,n}$ in order to solve the power optimization problem efficiently.

Approximation of the objective function (24): Let us start with objective function (24) by expanding it as follows:

$$R_u = B^M \underbrace{\sum_{m=1}^{M} \sum_{n=1}^{N^M} \epsilon^M_{mu,n} \log_2 \left(1 + \frac{P^M_{mu,n} h^M_{mu,n}}{\mathcal{I}^M_u + \mathcal{N}_0 B^M} \right)}_{R^M_{mu,n}}$$

$$+ B^L \underbrace{\sum_{l=1}^{L} \sum_{n=1}^{N^L} \epsilon^L_{lu,n} \log_2 \left(1 + \frac{P^L_{lu,n} h^L_{lu,n}}{\mathcal{N}_0 B^L} \right)}_{R^L_{lu,n}} + B^0 \underbrace{\sum_{n=1}^{N^0} \epsilon^0_{0u,n} \log_2 \left(1 + \frac{P^0_{0u,n} h^0_{0u,n}}{\mathcal{N}_0 B^0} \right)}_{R^0_{0u,n}}. \tag{27}$$

In (27), since we are maximizing the utility, then we need the R_u to be concave function. Note that the second and third terms (i.e., $R^L_{lu,n}$ and $R^0_{0u,n}$) are concave functions with respect to $P^L_{lu,n}$ and $P^0_{0u,n}$. In order to convert the objective function to a concave function, the first term (i.e., $R^M_{mu,n}$) needs to be concave with respect to $P^M_{mu,n}$. Let us start by expanding the $R^M_{mu,n}$ as follows:

$$R_{mu,n}^M = \underbrace{\epsilon_{mu,n}^M B^S \log_2 \left(\mathcal{N}_0 B^S + \sum_{m=1}^{M} \sum_{u=1}^{U} P_{mu,n}^M h_{mu,n}^M \right)}_{\hat{R}_{mu,n}^M}$$

$$\underbrace{-\epsilon_{mu,n}^M B^S \log_2 \left(\mathcal{N}_0 B^S + \sum_{\substack{\tilde{m}=1 \\ \tilde{m} \neq l}}^{M} \sum_{\substack{\tilde{u}=1 \\ \tilde{u} \neq u}}^{U} \epsilon_{\tilde{m}\tilde{u},n}^M P_{\tilde{m}\tilde{u},n}^M h_{\tilde{m}\tilde{u},n}^M \right)}_{\check{R}_{mu,n}^M}.$$

(28)

Note that $\hat{R}_{mu,n}^M$ is concave, because the log of an affine function is concave function [48]. Also, $\check{R}_{mu,n}^M$ is a convex function, and thus, it needs to be converted to a concave function. To tackle the nonconcavity of $\check{R}_{mu,n}^M$, successive convex approximation (SCA) method can be applied where in each iteration, the original function is approximated by a more tractable function at a given local point as given in Algorithm 1. Recall that $\check{R}_{mu,n}^M$ is convex in $P_{\tilde{m}\tilde{u},n}^M$, and any convex function can be globally lower-bounded by its first-order Taylor expansion at any point. Therefore, given $P_{\tilde{m}\tilde{u},n}^M(r)$ in iteration r, we obtain the following lower bound for $\check{R}_{mu,n}^M(r)$:

$$\check{R}_{mu,n}^M(r) \geq -\epsilon_{mu,n}^M B^S \log_2(\psi(r)) - \frac{\epsilon_{\tilde{m}\tilde{u},n}^M h_{\tilde{m}\tilde{u},n}^M}{\ln(2)\psi(r)} (P_{\tilde{m}\tilde{u},n}^M - P_{\tilde{l}\tilde{u},n}^M(r)),$$

(29)

where $\psi(r) = \mathcal{N}_0 B^S + \sum_{\substack{\tilde{m}=1 \\ \tilde{m} \neq l}}^{M} \sum_{\substack{\tilde{u}=1 \\ \tilde{u} \neq u}}^{U} \epsilon_{\tilde{m}\tilde{u},n}^M P_{\tilde{m}\tilde{u},n}^M h_{\tilde{m}\tilde{u},n}^M$. Now it can be seen that the objective function in (27) is a concave function in terms of $P_{su,n}^S$.

ALGORITHM 1 SCA Algorithm

Initial state: selecting a feasible set of $P_{\tilde{m}\tilde{u},n}^M$.

r=1.

repeat

Solving the formulated optimization problem in (24)–(26) using the interior point algorithm to find the next iteration approximated solution of $P_{\tilde{m}\tilde{u},n}^M(r)$.

until Reaching convergence.

Approximation of constraint (26): We can approximate this constraint by ensuring that the data rate from HAP l to any associated users u is smaller that the average backhaul constraint. In other words, constraint (16) can be approximated as:

$$\sum_{n=1}^{N^L} \epsilon_{lu,n}^L \left(1 + \frac{P_{lu,n}^L h_{lu,n}^L}{\mathcal{N}_0 B^S} \right) \leq \frac{\omega}{\sum_{n=1}^{N^L} \epsilon_{lu,n}^L}, \quad \forall l, \forall u, \tag{30}$$

where $\omega = 2^{\left(\sum_{w=0}^{W} \bar{R}_{wl}\right)/B^S}$.

Now, our power allocation optimization problem given in (24)–(26) becomes convex problem with respect to $P_{su,n}^S$ for fix associations and HAPs' locations. Thus, the duality gap of our convex power allocation problem is zero. Hence, We can solve our convex transmit power allocation optimization problem by exploiting its strong duality.

4.1.2 Low complexity solution

This subsection proposes an efficient and low complexity solution that mitigates the intercell interference between users connected to different terrestrial stations within the ground-tier. We group the users into two groups, (1) center users: the users located close to the center of the terrestrial stations' coverage and (2) edge users: users located at the edges of the terrestrial stations' coverage. Let us consider that terrestrial stations can be connected with the center users only. In this case, the intercell interference is assumed to be $\mathcal{I}_{mu}^M = 0$ due to large distances between users connected to different terrestrial stations and using the same resource allocations within the ground-tier. In counterpart, the edge users can be served by the HAPs or satellite station as shown in Fig 4. Therefore, (24) converted to a convex function with respect to $P_{su,n}^S$.

As a result, the formulated optimization problem in (14)–(16) becomes a linear assignment problem with respect to $\epsilon_{su,n}^S$ and δ_{wl}. Hence, it can be solved efficiently by using the Hungarian algorithm with complexity equal to $(MN^M+LN^L+N^0)^3$ [49].

Apart from that, by only approximating (26), the formulated optimization problem given in (24)–(26) becomes a convex optimization problem in terms of $P_{su,n}^S$. Hence, it can be solved efficiently by solving its dual problem.

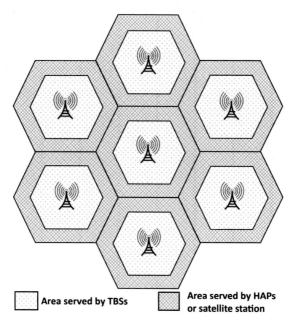

Area served by TBSs   Area served by HAPs or satellite station

Fig. 4 Applying frequency partitioning technique to the ground-tier.

4.2 Long-term stage

Even by fixing the access/backhaul associations and stations' transmit power, the formulated optimization given in (14)–(20) is still nonconvexity optimization problem. Therefore, we propose an efficient and low complexity solution based on shrink and realign algorithm. This algorithm has several advantages over similar heuristic algorithms investigated in the literature such as simpler implementations, lower complexity, and quicker convergence to a near optimal solution. Moreover, because we solve this optimization problem for long-term stage, where the time slots are relatively very large compared to the channel coherence time, we then focus on the average statistics of the network, such as average channel gains and average user distributions. This implies that average throughput of the users is optimized instead of the instantaneous throughput.

The proposed algorithm starts by forming an initial population set of next position candidates T_l, $\forall l$ in a spherical shape with radius $r(i)$ around all current HAPs' locations. We then solve the average associations and stations' transmit powers by finding the maximum number of average users inside the HAPs coverage areas for each candidates combination. Meaning, we determine the value of $U_{\max}^L = \sum_{l=1}^{L} \sum_{n=1}^{N^S} \sum_{u=1}^{U} \epsilon_{lu,n}^L$, after excluding the users connected to the terrestrial stations. Then, we find the initial best local candidate combinations $T^{i,\text{local}} = t_l^{i,\text{local}}, \forall l$ that provides the maximum number of users associated with HAPs for iteration i. After that, shrink and realign process is applied recursively to find the best global solution $T^* = t_l^*, \forall l$ by generating a new candidates on an updated sphere radius $(r(i+1) = r(i)/2)$ around each local solution. This process is repeated until the size of the sample space decreases below a certain threshold $I_{\max}$ or no improvement can be made. Fig. 5 shows an example of using our algorithm in 2D using two HAPs ($L = 2$), $T = 4$, and with three maximum iterations $I_{\max} = 3$. The details of the location of the HAPs algorithm that optimizes the locations of the HAPs based on average users' locations are given in Algorithm 2.

5. Simulation and numerical results

In this section, selected numerical results are provided to illustrate the advantages of our proposed integrated model. We run the numerical results with a total area of 180 km × 180 km. Inside this area, we assume we have

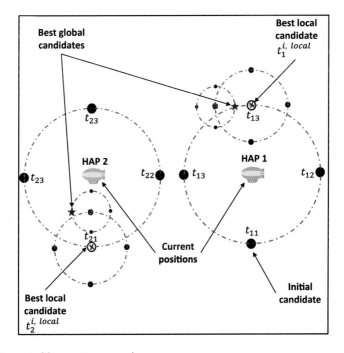

Fig. 5 Proposed heuristic approach.

ALGORITHM 2 Optimizing the location of HAPs
1: i=1.
2: Generating an initial set of candidates T in a spherical shape of radius $r = r(i)$
    around each HAP $J_l^c(t_l)$, $l = 1...L$.
3: **while** Not converging or reaching l_{max} **do**
4:     **for** $l = 1...L$ **do**
5:        **for** $t_l = 1...T$ **do**
6:           Finding the best value of $\epsilon_{su,n}^S$, δ_{wl}, and $P_{su,n}^S$ as explained in Section. IV
    using average users statistics.
7:           Computing $U_{max}^L = \sum_{l=1}^L \sum_{n=1}^{N^S} \sum_{u=1}^U \epsilon_{lu,n}^L$.
8:        **end for**
9:     **end for**
10:    Finding $t_l^{i,local} = \underset{t_t}{\arg\max} \, U_{max}^L$, (i.e., $t_l^{i,local}$ that indicates the best local

    candidate index that results in the highest U_{max}^L for iteration i).
11:    $r = r(i)/2$
12:    Applying the shrink and realign process for each of the local solution.
13:    Updating the iteration i=i+1.
14: **end while**

U ground users that are distributed in three different areas (i.e., area 1: 70 km × 70 km, area 2: 70 km × 70 km, and area 3: the remaining) with different users' density distributions. Area 1 contains 30 GBSs with x range of (55–125 km) and y range of (0–70 km) and consists 40% of total ground users' number. Subarea 2 has no GBSs with x range of (55–125 km) and y range of (110–180 km) and consists 30% of total ground users' number. Note that the total number of HAPs used are eight. Table 3 summarizes the simulations targeted areas. The satellite station is assumed to be in a fixed location with the coordinate [90, 90, 2000] km. Also, we consider $L = 5$ HAPs initially flying at locations $\mathbf{J}_1^L = [90, 90, 18]$ km, $\mathbf{J}_1^L = [30, 30, 18]$ km, $\mathbf{J}_1^L = [150, 30, 18]$ km, $\mathbf{J}_1^L = [30, 150, 18]$ km, $\mathbf{J}_1^L = [150, 150, 18]$ km. We assume that $W = 4$ and located in the four corners of the desired area. In our simulations, we assume $N^M = 50$, $N^L = 100$, and $N^0 = 200$ available resource blocks in the access link. The maximum transmit power of terrestrial station, HAP, and satellite station are, respectively, given as {40,100,250} W. The noise power $\mathcal{N}_0$ is assumed to be − 174 dBm. In

Table 3 Numerical scenarios.

	Area 1	Area 2	Area 3
Area (km × km)	70 × 70	70 × 70	Remaining
Dimensions [X],[Y] (km)	[55–125], [0–70]	[55–125], [110–180]	Remaining
Users distribution	40%	30%	30%
Example area	Urban	After hurricane	Remote/ rural

Table 4 Simulation parameters.

Constant	Value	Constant	Value	Constant	Value
f_c^M (GHz)	1.8	f_c^L (GHz)	3.0	f_c^0 (GHz)	5.0
f_c (GHz)	3.4	χ	2	κ_L, κ_0	10
B^M (kHz)	1.8	B^L (MHz)	1.0	B^L (MHz)	2.0
Ω_0	0.372	Ω_1	0.0129	Ω_2	7.64

Table 4, we present the values of the remaining environmental parameters used in the simulations unless otherwise stated [33,34,42].

Fig. 6 plots the optimization of HAPs' locations and backhaul associations using average users location for two high and low users densities with $P_0 = 40$ W, $B_0 = 4$ MHz, and $\overline{P}_l = 100$ W. The figure investigate the backhaul association of HAPs with the gateways. For instance, Fig. 6A shows that the HAPs try to accommodate maximum number of users by finding the best HAPs' locations that improved the access and backhaul links at the same time. Note that the users located outside the coverage areas of HAPs and terrestrial stations are associated with satellite station. While, in the case of low dense users, HAPs try to associated with all users as shown in Fig. 6B. This is because associating with HAPs, in general, provide a better data rate rater than associating to satellite station due to the high path-loss attenuation between satellite station and ground users. In this case, the HAPs can provide services to ground users with much lower delay and higher throughput.

Fig. 7 plots the average data rate per user (i.e., $\sum_{u=1}^{U} U$) vs total number of users in the three subareas with $P_0 = 40$ W, $B_0 = 4$ MHz, and $\overline{P}_l = 100$ W. Our proposed solutions (i.e., approximated and low complexity

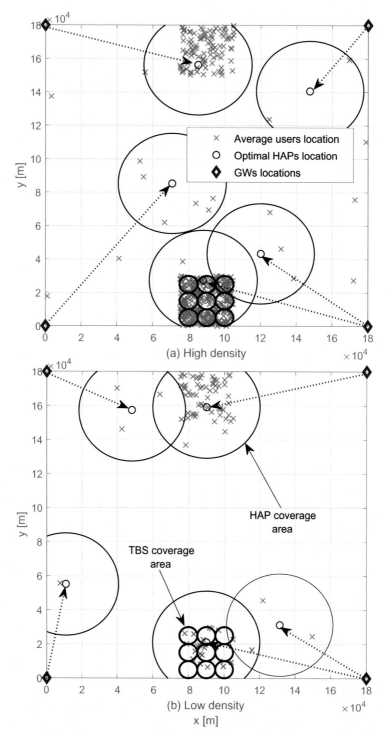

Fig. 6 HAP placement optimization with backhaul associations.

solutions) are compared with two benchmark solutions: 1—optimizing the access and backhaul associations in addition to the HAPs' locations with uniform power distribution (i.e., $P_{su,n}^S = \overline{P}_S/N^S$), and 2—optimizing only the placement of the HAPs using random access and backhaul associations with uniform power distribution. The figure shows that for fixed resources, as U increases, the average data rate decreases due to the limitation of the available resources. Furthermore, the figure shows that the approximate solution and low complexity solution achieve almost the same performance for low values of U. However, there will be a gap between the two proposed solutions when U is relatively large. This is can be explained because the low complexity solution forces some HAPs to keep covering certain areas under the terrestrial stations coverage areas to implement the FP technique as discussed in Section 4.1.2 and shown in Fig. 4. This will not affect the performance when U is relatively low, but when U is large, instead of keeping HAPs covering the terrestrial stations areas maybe be better to move them to different locations for a better total coverage/performance. Thanks to the HAPs, because HAPs have large coverage areas, this gap for large U is still acceptable. This figure also shows that our proposed solutions outperform the other two benchmarks solutions. For instance, using $U = 400$, our proposed solution can enhance the average rate throughput by at least 39% and 88% compared to optimize the associations with uniform power and to random associations with uniform power, respectively. Furthermore, it can be noticed that the gap between the proposed solutions and benchmarks solutions is increased as U increases. This is because as U increases, the need of managing and optimizing the power is needed more.

Additionally, all simulations show that MSU leads to the highest average data rate in the system. However, this comes at the expense of fairness as it is shown in Table 5. Indeed, the table compares between the two different utilities for the same channel realization with fixed $P_0 = 40$ W, $B_0 = 4$ MHz, $\overline{P}_l = 100$ W, and $U = 400$ (for this realization, 160 user associated to terrestrial stations, 172 user associated to HAPs 172, and 68 user associated to satellite station). In Table 5, we denote $\overline{R}^S$, R_{max}^S, and R_{min}^S as average rate, maximum rate, and minimum rate in tier S, respectively. Also, P_{max}^S and P_{min}^S denoted as maximum and minimum transmit powers in tier S. By using one realization, it can be shown that MSU allocates most of the resources to users having the best channel conditions. On the other hand, MMU approach maximizes the minimum data rate and thus provides almost the same rate for all users in each tier. Hence, MMU leads to more fairness performance. The choice of the utility is related to the service given

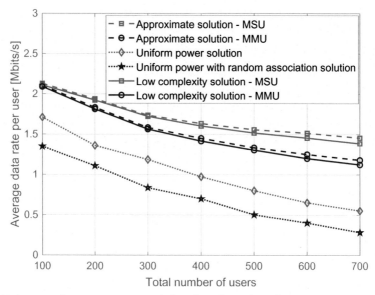

Fig. 7 Average data rate per user as a function of number of users.

Table 5 Comparison between MSU and MMU for fixed $P_0 = 40$ W, $B_0 = 4$ MHz, $\bar{P}_l = 100$ W, and $U = 400$.

	Users associated with	$\bar{R}^S$ Mbits/s	(R^S_{max}, P^S_{max}) (Mbits/s,W)	(R^S_{min}, P^S_{min}) (Mbits/s,W)
MSU	Terrestrial stations	2.14	(4.13, 15.25)	$(\sim 0, \sim 0)$
	HAPs	1.92	(3.19, 39.70)	(.0415, 1.2)
	Satellite	(0.016)	(0.085, 97.53)	$(\sim 0, \sim 0)$
MMU	Terrestrial stations	(1.91)	(1.96, 5.49)	(1.88, 8.20)
	HAPs	(1.50)	(1.81, 3.91)	(1.39, 4.5)
	Satellite	(0.01)	(0.01, 4.81)	(0.01, 4.65)

to the users. For instance, if the application requires same downlink rates between users, then MMU can be used. On the other hand, if it consists in a pure transmission without priorities, then MSU could be employed.

Fig. 8 illustrates the performance of the average data rate for users associated with HAPs using $U = 400$. More specifically, Fig. 8 plots the effect of the HAPs transmit power on the average data rate of HAPs' users for different backhaul transmit powers $P_0 = \{10, 20, 30, 40\}$ W with

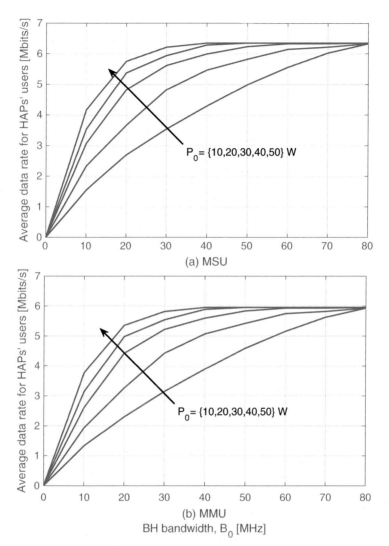

Fig. 8 Average data rate of HAPs' users vs HAPs' peak power.

$\overline{P}_l = 100$ W. This figure shows that the average data rate of HAPs' users is improving with as B_0 increases up to a certain cutoff value, that depends on the backhaul rate constraint as given in Eq. (16). This is because starting from this cutoff value, the average data rate can not be enhanced further because it depends on the value of $P_{lu,n}^L$ which is limited by $\overline{P}_L$ as given in (15). Also, Fig. 8 shows that the cutoff value depends on the backhaul transmit

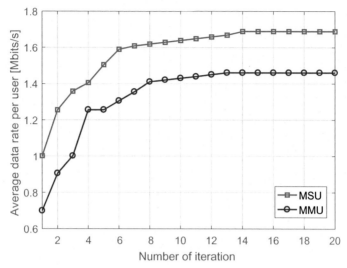

Fig. 9 Convergence of the proposed SR algorithm.

power P_0. As P_0 increases the corresponding cutoff value decreases in order to respect constraint (16). In this case, the bottleneck is the access rate constraint which is limited by $\overline{P}_L$.

In Fig. 9, we plot the convergence speed of the SR algorithm, defined by the number of iterations needed to reach convergence, for both utilities, MSU and MMU. Note that one iteration in Fig. 9 corresponds to one iteration of the "while loop" given in Algorithm 2 line 4–13. It can be noticed that the algorithm is converge within around 8–13 iterations.

6. Conclusion and future directions

6.1 Conclusion

In this book chapter we proposed an efficient optimization framework using of ground–air–space integration to provide downlink connectivity to the ground users taking into consideration the access and backhaul limitations. The objective was to maximize the users' throughput by optimizing the access and backhaul associations, transmit powers of the stations, and the HAPs' locations. We proposed an approximate and a low complexity solutions to optimally determine the decision variables. The simulation results illustrated the behavior of our approach and their significant impacts on the users' throughput using different utility metrics.

6.2 Future direction: Hybrid RF/FSO in backhaul

Prospective demands of next-generation wireless networks are ambitious and will be required to support data rates 1000 times higher and round-trip latency 10 times lower than current wireless networks. The RF spectrum is expected to become more congested for emerging applications in next-generation wireless communications and eventually insufficient to accommodate the increasing demand of mobile data traffic. Relying on improved RF access in legacy bands alone is not sufficient, thus, it is critical to embrace a comprehensive solution with high spectral reuse by supplementing RF with other emerging wireless technologies in directional high-frequency bands [50]. Free-space optical (FSO) communications present a promising complementary solution not only to meet the exploding demand for wireless communications but also to be used in providing high backhaul data rates for long distances in the case of infrastructure-less situations.

Besides the wide available bandwidth in Terra-Hertz without any restriction or regulation (license-free spectrum), FSO communications hold the advantages of ease of deployment, high transmission security, immunity to electromagnetic interference due to the very narrow laser beams along with lesser size, weight, and power compared to their RF counterparts, and as a consequence they are well-suited for inter-HAPs, ground station-to-HAP, and HAP-to ground station communications [22]. There is limited research in the literature that has proposed equipping HAPs with FSO transceivers [22,51]. For instance, in [22], the authors provide an overview of HAPs equipped with optical transceivers and show that by using laser beams, a data rate of several Gbps can be achieved. In [51], closed form expressions for bit-error-rate and average capacity are derived for multihop FSO links in the stratosphere. However, all these works have not considered managing the resource allocation in satellite–airborne–terrestrial network integration. In addition, the previous works have not considered issues surrounding access and backhaul links communications.

Why equipping HAPs with free-space-optical (FSO) transceivers is a game changer:

Some large geographical areas may require HAP stations to provide downlink wireless service, but the service is limited by the backhaul bottleneck. For example, in 2017, Hurricane Irma damaged a significant proportions of the wireless cell towers in several parts of Florida. Even after 2 days following the hurricane, Irma's wrath had caused more than 50% cell tower failure in some counties in Miami, FL [52]. In such circumstances, HAPs are capable of reaching such affected areas thanks to their quick and dynamic deployment. The success of deploying HAPs in remote or challenging areas does not depend only on their integration with ground users through the access links, but also on other parameters related to the backhaul links. Therefore, the integrating the HAPs FSO transceivers can be a great way to solve wireless connectivity in challenging areas.

In our future direction, we propose to use the RF/FSO link for backhaul links only as shown in Fig. 10. The RF and FSO choice will depend on several factors such as environmental/weather conditions and the feasibility of LoS. Meanwhile, the RF band can be used in the access link due to the difficulties in tracking the movement of ground users and maintaining the LoS.

To illustrate the RF/FSO backhaul bottleneck, Fig. 11 plots the average downlink data rate vs the HAPs' peak power. The figure shows that as the HAPs' peak power increases, the downlink data rate rises to a certain value. This is because, by starting from this HAP peak power, the average data rate cannot be improved because it also depends on the backhaul data rate from HAPs to gateways or HAPs to satellite links. Also, note that the average data rate increases as the backhaul bandwidth increases because increasing the backhaul bandwidth also increases the value of the backhaul data rate, thus increasing the backhaul bottleneck. Therefore, hybrid RF/FSO communication links can be used to mitigate backhaul bottleneck limitations and thus enhance performance.

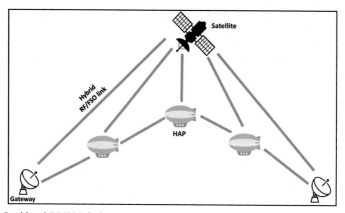

Fig. 10 Backhaul RF/FSO link.

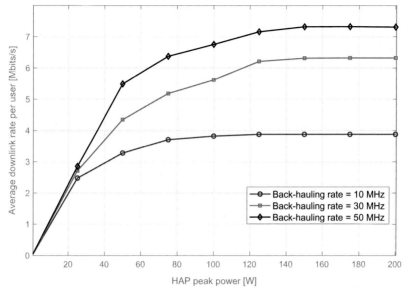

Fig. 11 Average downlink data rate vs HAPs' peak power.

6.3 Future direction: HAP and tethered balloons integration

On another front, achieving good uplink data transmission becomes necessarily for the two way communications. Some large areas may require HAPs to provide downlink and uplink wireless services where the power limitations of the ground users require intermediate nodes, such as relays, which broadcast and/or amplify downlink and uplink signals. Relays not only improve these signals but also provide a new dimension to next-generation wireless networking and service provisioning. In such circumstances, HAPs and relays can reach such affected areas thanks to their quick, dynamic deployment. The success of deploying tethered balloons as relays in remote or challenging areas does not depend only on their integration with ground users through access links but also on other parameters related to backhaul links. Therefore, the integration of HAP and tethered balloons technology with highly efficient placement and resource management wonderfully generates wireless connectivity in challenging areas. In this study, we propose tethered balloons powered by renewable energy (RE) sources as downlink and uplink relays to serve users in remote areas. We consider this a plausible proposal since in remote areas the tethered balloons would serve fewer users and would therefore consume less energy.

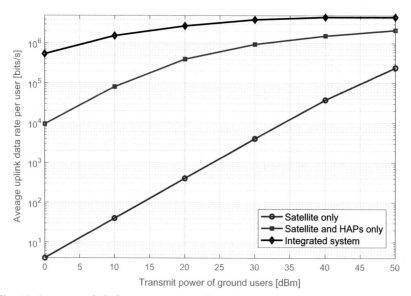

Fig. 12 Average uplink data rate vs ground user's transmit power.

Fig. 12 illustrates the average uplink data rate per ground user (i.e., the total sum rate divided by the number of users) vs the transmit power of ground users. We compare the integrated system of satellite, HAPs, and tethered balloons with two benchmark solutions: (1) using only a satellite station for the uplink transmission and (2) using both satellites and HAPs for the uplink transmission. The figure shows that the integrated system achieves a much better uplink data rate than the two benchmarks. For instance, using 20 dBm transmission power, the achievable uplink data rate can be enhanced from 0.3 Kbits/s and 0.3 Mbits/s using satellite only and both satellite and HAPs, respectively, to more than 2 Mbits/s using the proposed integrated system. This is because tethered balloons can work as relays to mitigate path loss and other unwanted effects. Furthermore, the figure shows that the proposed integrated system increases as the transmit power rises to a certain value (around 40 dBm). After this value, the performance remains constant. This is because of the limitation in the relay link from tethered balloons to HAPs.

In the conclusion, equipping HAPs with FSO transceivers can enhance the network performance in general, however, this will add more complexity to the problem by optimizing extra parameters such as the LoS angles alignment.

References

[1] E.C. Strinatia, S. Barbarossa, J.L. Gonzalez-Jimenez, D. Kténas, N. Cassiau, C. Dehos, 6G: The Next Frontier, 2019. https://arxiv.org/abs/1901.03239.

[2] F. Tariq, M. Khandaker, K.-K. Wong, M. Imran, M. Bennis, M. Debbah, A Speculative Study on 6G, 2019. https://arxiv.org/abs/1902.06700.

[3] M.B. Walid Saad, M. Chen, A vision of 6G wireless systems: applications, trends, technologies, and open research problems, 2019. https://arxiv.org/abs/1902.10265.

[4] J. Liu, Y. Shi, Z.M. Fadlullah, N. Kato, Space-air-ground integrated network: a survey, IEEE Commun. Surv. Tutorials 20 (4) (2018) 2714–2741, https://doi.org/10.1109/COMST.2018.2841996.

[5] C. Chen, A. Grier, M. Malfa, E. Booen, H. Harding, C. Xia, M. Hunwardsen, J. Demers, K. Kudinov, G. Mak, B. Smith, A. Sahasrabudhe, F. Patawaran, T. Wang, A. Wang, C. Zhao, D. Leang, J. Gin, M. Lewis, D. Nguyen, K. Quirk, High-speed optical links for UAV applications, in: SPIE Free-Space Laser Communication and Atmospheric Propagation XXIX, March, vol. 10096, 2017, https://doi.org/10.1117/12.2256248.

[6] K. An, T. Liang, G. Zheng, X. Yan, Y. Li, S. Chatzinotas, Performance limits of cognitive-uplink FSS and terrestrial FS for Ka-band, IEEE Trans. Aerospace Electron. Syst. 55 (5) (2019) 2604–2611.

[7] R. Musumpuka, T.M. Walingo, J.M. Smith, Performance analysis of correlated handover service in LEO mobile satellite systems, IEEE Commun. Lett. 20 (11) (2016) 2213–2216.

[8] R.M. Calvo, J. Poliak, J. Surof, A. Reeves, M. Richerzhagen, H.F. Kelemu, R. Barrios, C. Carrizo, R. Wolf, F. Rein, A. Dochhan, K. Saucke, W. Luetke, Optical technologies for very high throughput satellite communications, in: SPIE Free-Space Laser Communications XXXI, March, vol. 10910, 2019, https://doi.org/10.1117/12.2513819.

[9] Starlink, 2018. https://www.starlink.com/.

[10] OneWeb, Revolutionizing the economics of space, 2021 (Available online) https://onewebsatellites.com/.

[11] Telesat, 2018. https://www.telesat.com/services/leo/.

[12] S. Adibi, A. Mobasher, T. Tofigh, Fourth-Generation Wireless Networks: Applications and Innovations, Information Science Publishing, 2009.

[13] X. Yan, H. Xiao, C. Wang, K. An, Outage performance of NOMA-based hybrid satellite-terrestrial relay networks, IEEE Wireless Commun. Lett. 7 (4) (2018) 538–541.

[14] K. An, M. Lin, T. Liang, J. Wang, J. Wang, Y. Huang, A.L. Swindlehurst, Performance analysis of multi-antenna hybrid satellite-terrestrial relay networks in the presence of interference, IEEE Trans. Commun. 63 (11) (2015) 4390–4404.

[15] U. Park, H.W. Kim, D.S. Oh, B.J. Ku, Flexible bandwidth allocation scheme based on traffic demands and channel conditions for multi-beam satellite systems, in: Proc. of the IEEE Vehicular Technology Conference (VTC Fall), Quebec City, QC, Canada, September, 2012, pp. 1–5.

[16] J. Lei, M.A. Vazquez-Castro, Joint power and carrier allocation for the multibeam satellite downlink with individual SINR constraints, in: Proc. of the IEEE International Conference on Communications (ICC), Cape Town, South Africa, May, 2010, pp. 1–5, https://doi.org/10.1109/ICC.2010.5502506.

[17] F. Li, K. Lam, J. Hua, K. Zhao, N. Zhao, L. Wang, Improving spectrum management for satellite communication systems with hunger marketing, IEEE Wireless Commun. Lett. 8 (3) (2019) 797–800.

[18] N. Zhang, S. Zhang, P. Yang, O. Alhussein, W. Zhuang, X.S. Shen, Software defined space-air-ground integrated vehicular networks: challenges and solutions, IEEE Commun. Mag. 55 (7) (2017) 101–109, https://doi.org/10.1109/MCOM.2017.1601156.

[19] A. Mohammed, A. Mehmood, F. Pavlidou, M. Mohorcic, The role of high-altitude platforms (HAPs) in the global wireless connectivity, Proc. IEEE 99 (11) (2011) 1939–1953.

[20] S. Karapantazis, F. Pavlidou, Broadband communications via high-altitude platforms: a survey, IEEE Commun. Surv. Tutorials 7 (1) (2005) 2–31.

[21] X. Cao, P. Yang, M. Alzenad, X. Xi, D. Wu, H. Yanikomeroglu, Airborne communication networks: a survey, IEEE J. Selec. Areas Commun. 36 (9) (2018) 1907–1926.

[22] F. Fidler, M. Knapek, J. Horwath, W.R. Leeb, Optical communications for high-altitude platforms, IEEE J. Selec. Top. Quantum Electron. 16 (5) (2010) 1058–1070, https://doi.org/10.1109/JSTQE.2010.2047382.

[23] M. Sharma, D. Chadha, V. Chandra, High-altitude platform for free-space optical communication: performance evaluation and reliability analysis, IEEE/OSA J. Opt. Commun. Networking 1943-0620, 8 (8) (2016) 600–609, https://doi.org/10.1364/JOCN.8.000600.

[24] J. Marriott, B. Tezel, Z. Liu, N. Stier, Trajectory optimization of solar-powered high-altitude long endurance aircraft, Facebook Res. (2018).

[25] H. Bolandhemmat, B. Thomsen, J. Marriott, Energy-Optimized Trajectory Planning for High Altitude Long Endurance (HALE) Aircraft, in: Proc. of the 18th European Control Conference (ECC), Naples, Italy, Aug, 2019, pp. 1486–1493.

[26] H. Lu, Y. Gui, X. Jiang, F. Wu, C.W. Chen, Compressed robust transmission for remote sensing services in space information networks, IEEE Wireless Commun. 26 (2) (2019) 46–54.

[27] Z. Yang, A. Mohammed, Wireless communications from high altitude platforms: applications, deployment, and development, in: Proc. of the 12th IEEE International Conference on Communication Technology, Nanjing, China, 2010, pp. 1476–1479.

[28] Aurora Flight Sciences, A Boeing Company, ODYSSEUS Project, in: ODYSSEUS Project, https://www.aurora.aero/odysseus-high-altitude-pseudo-satellite-haps/.

[29] Ordnance Survey, Astigan High Altitude Pseudo Satellite Project, https://www.ordnancesurvey.co.uk/news/astigan-pseudo-satellite.

[30] Airbus, Zephyr, https://www.airbus.com/defence/uav/zephyr.html.

[31] AeroVironment, AeroVironment Avinc HAPS project, https://www.avinc.com/about/haps.

[32] Google, Loon Project, https://loon.com/.

[33] R. Zong, X. Gao, X. Wang, L. Zongting, Deployment of high altitude platforms network: a game theoretic approach, in: Proc. of the International Conference on Computing, Networking and Communications (ICNC), Maui, HI, USA, 2012, pp. 304–308.

[34] A. Ibrahim, A.S. Alfa, Using Lagrangian relaxation for radio resource allocation in high altitude platforms, IEEE Trans. Wireless Commun. 14 (10) (2015) 5823–5835.

[35] D. Grace, J. Thornton, Guanhua Chen, G.P. White, T.C. Tozer, Improving the system capacity of broadband services using multiple high-altitude platforms, IEEE Trans. Wireless Commun. 4 (2) (2005) 700–709.

[36] J. Tong, Y. Lu, D. Zhang, G. Cui, W. Wang, Max-min analog beamforming for high altitude platforms communication systems, in: Proc. of the IEEE International Conference on Communications in China (ICCC), Beijing, China, August, 2018, pp. 872–876.

[37] Z. Chen, T.B. Lim, M. Motani, HAP-assisted LEO satellite downlink transmission: an energy harvesting perspective, in: Proc. of the 15th IEEE International Workshop on Signal Processing Advances in Wireless Communications (SPAWC), Toronto, ON, Canada, June, 2014, pp. 194–198.

[38] S. Zhou, G. Wang, S. Zhang, Z. Niu, X.S. Shen, Bidirectional mission offloading for agile space-air-ground integrated networks, IEEE Wireless Commun. 26 (2) (2019) 38–45.

[39] 3rd Generation Partnership Project, Evolved Universal Terrestrial Radio Access E-UTRA; Physical channels and modulation, 2013. 3GPP TS 36.211 3GPP V 11.2.0, Release 11, April.

[40] D.A.J. Pearce, D. Grace, Optimum antenna configurations for millimetre-wave communications from high-altitude platforms, IET Communications 1 (3) (2007) 359–364.

[41] E. Falletti, M. Laddomada, M. Mondin, F. Sellone, Integrated services from high-altitude platforms: a flexible communication system, IEEE Commun. Mag. 44 (2) (2006) 85–94.

[42] A. Abdi, W.C. Lau, M.-S. Alouini, M. Kaveh, A new simple model for land mobile satellite channels: first- and second-order statistics, IEEE Trans. Wireless Commun. 2 (3) (2003) 519–528, https://doi.org/10.1109/TWC.2003.811182.

[43] A. Iqbal, K.M. Ahmed, Outage probability analysis of multi-hop cooperative satellite-terrestrial network, in: Proc. of the 8th Electrical Engineering/ Electronics, Computer, Telecommunications and Information Technology (ECTI) Association, Phuket, Thailand, May, 2011, pp. 256–259.

[44] F. Dong, H. Li, X. Gong, Q. Liu, J. Wang, Energy-efficient transmissions for remote wireless sensor networks: an integrated HAP/satellite architecture for emergency scenarios, Sensors 15 (9) (2015) 22266–22290.

[45] X. Bi, J. Zhang, Y. Wang, P. Viswanath, Fairness improvement of maximum C/I scheduler by dumb antennas in slow fading channel, in: Proc. of the 72nd IEEE Vehicular Technology Conference (VTC Fall), Ottawa, Ontario, Canada, 2010, pp. 1–4.

[46] G. Song, Y. Li, Cross-layer optimization for OFDM wireless networks—Part I: Theoretical framework, IEEE Trans. Wireless Commun. 4 (2) (2005) 614–624.

[47] Gurobi Optimizer, Gurobi Optimizer Reference Manual, 2016. Available: http://www.gurobi.com/. (online).

[48] S. Boyd, L. Vandenberghe, Convex Optimization, Cambridge University Press, New York, NY, USA, 2004, ISBN: 0521833787.

[49] H.W. Kuhn, The Hungarian Method for the Assignment Problem, in: 50 Years of Integer Programming 1958-2008, Springer, Berlin, Heidelberg, 2010 (Chapter 2).

[50] A. Sevincer, A. Bhattarai, M. Bilgi, M. Yuksel, N. Pala, LIGHTNETs: smart LIGHTing and mobile optical wireless networks—a survey, IEEE Commun. Surv. Tutorials 15 (4) (2013) 1620–1641.

[51] M. Sharma, D. Chadha, V. Chandra, High-altitude platform for free-space optical communication: performance evaluation and reliability analysis, IEEE/OSA J. Opt. Commun. Networking 8 (8) (2016) 600–609.

[52] Federal Communications Commission, Communications Status Report for Areas Impacted by Hurricane Irma September 11, 2017, 2017. FCC report.

About the authors

Ahmad Alsharoa was born in Irbid, Jordan. He received the Ph.D. degree with co-majors in electrical engineering and computer engineering from Iowa State University (ISU), Ames, IA, USA, in May 2017. He is currently an Assistant Professor with the Missouri University of Science and Technology (Missouri S&T). His current research interests include wireless networks, UAV and HAP communications, edge computing, optical communications, the Internet of Things (IoT), energy harvesting, and self-healing networks. He was a recipient of the International Scholarships Award, Research Excellence Award, and GPSS Research Award from ISU and Preeminent Postdoctoral Program (P3) Award from the University of Central Florida (UCF) in 2015, 2016, 2017, and 2018, respectively.

Emna Zedini was born in Beja, Tunisia. She received the Engineering and the M.Sc. degrees in Telecommunications from the École Supérieure des Communications de Tunis (SUP'COM), Tunis, Tunisia, in 2010 and 2011, respectively. She received the Ph.D. degree in Electrical Engineering from King Abdullah University of Science and Technology (KAUST), Saudi Arabia, in 2016. In 2017, she joined the college of science and engineering at Hamad Bin Khalifa University (HBKU), Doha, Qatar as a postdoctoral researcher. Her research interests include channel modeling and performance analysis of terrestrial and underwater optical wireless communication systems, with current research focusing on high throughput satellite systems based on optical feeder links.

Mohamed-Slim Alouini was born in Tunis, Tunisia. He received the Ph.D. degree in Electrical Engineering from Caltech in 1998. He served as a faculty member at the University of Minnesota then in the Texas A&M University at Qatar before joining in 2009 the King Abdullah University of Science and Technology (KAUST) where he is now a Distinguished Professor of Electrical and Computer Engineering. He is a Fellow of the IEEE and OSA and he is currently actively working on addressing the uneven global distribution, access to, and use of information and communication technologies in far-flung, low-density populations, low-income, and/or hard-to-reach areas.

CHAPTER TWO

Evaluating software testing techniques: A systematic mapping study

Mitchell Mayeda and Anneliese Andrews
University of Denver, Denver, CO, United States

Contents

Advances in Computers, Volume 123
ISSN 0065-2458
https://doi.org/10.1016/bs.adcom.2021.01.002

Abstract

Software testing techniques are crucial for detecting faults in software and reducing the risk of using it. As such, it is important that we have a good understanding of how to evaluate these techniques for their efficiency, scalability, applicability, and effectiveness at finding faults. This article enhances our understanding of software testing technique evaluations by providing an overview of the state of the art in research and structuring the field to assist researchers in locating types of evaluations they are interested in. To do so a systematic mapping study is performed. Three hundred and sixty-five primary studies are systematically collected from the field and each mapped into categories based on numerous classification schemes. This reveals the distribution of research by each category and identifies where there are research gaps. It also results in a mapping from each combination of categories to actual papers belonging to them; allowing researchers to very quickly locate all of the testing technique evaluation research with properties they are interested in. Further classifications are performed on case study and experiment evaluations in order to assess the relative quality of these evaluations. The distribution of research by various category combinations is presented along with a large table mapping each category combination to the papers belonging to them. We find a majority of evaluations are empirical evaluations in the form of case studies and experiments, most of them are of low quality based on proper methodology guidelines, and relatively few papers in the field discuss how testing techniques should be evaluated.

1. Introduction

Software testing is a vital process for detecting faults in software and reducing the risk of using it. With a rapidly expanding software industry and a heavy reliance on increasingly prevalent software, there is a serious demand for employing software testing techniques that are efficient, scalable, applicable, and effective at finding faults. Utilizing such testing techniques to reduce the risk of using software can help avoid catastrophes that jeopardize safety or cost companies millions of dollars, such as when Intel spent $475 million replacing processors due to inaccurate floating point number divisions [1]. Given the importance of applying high-quality software testing techniques, understanding how they should be evaluated is also crucial. What is the current state of the art in research evaluating software testing techniques and where are there gaps in research? As a researcher looking to evaluate a particular technique, how should I do so?

A systematic mapping study is a methodology that is useful for providing an overview of a research area by classifying reports in it and counting the number of them belonging to each category in the classification.

For example, one can classify papers in a field by their publication year with each year being a category in the classification. Counting the number of papers belonging to each category (in this case the number of papers published each year) can give us an idea of activity level in the field over time. Similarly, classifying papers based on their content gives us a sense of what content is commonly researched and where there are research gaps. Such classifications can also provide higher level insight regarding the current state of the art. As an example from this article, classifying papers by the method they utilized for evaluating software testing techniques gives a very general sense of which methods are commonly used for evaluations. Considering this classification with others such as the testing technique type or dimension of evaluation allows us to answer more interesting questions about the state of the art: What evaluation method is most commonly used when evaluating the efficiency of mutation testing techniques? What is the distribution of evaluation methods when evaluating the effectiveness of white-box testing techniques? Additionally, classifications can be used to construct a mapping from categories to sets of papers belonging to them; allowing researchers to very easily locate papers in the field belonging to categories they are interested in. Here, we utilize a systematic mapping study in the field of research evaluating software testing techniques to achieve our main goals of (1) summarizing recent publication trends and (2) identifying research gaps and the state of the art when it comes to evaluating software testing techniques. We hope by structuring the field that we can provide guidance to other researchers who are unsure of how to evaluate their particular testing technique and point them to specific papers that have evaluated similar techniques. We also hope that we can provide direction for future work and initiate improvements in areas where evaluations are of lower relative quality. Our systematic mapping process follows guidelines proposed by Petersen et al. [2] and is discussed in more detail in Section 3.

This article is organized as follows. Section 2 provides a background on other surveys and mapping studies. Section 3 provides an overview of the systematic mapping process and a detailed explanation of each step in the process as it relates to our particular mapping study. Section 4 presents the study classification schemes used to classify articles into categories for this study. Section 5 presents the results of the data mapping. Section 6 provides a discussion of the results. Section 7 demonstrates the use of the resulting map with a case study. Finally, Section 8 considers threats to the validity of our findings followed by a conclusion and future work in Section 9.

2. Background

Other relevant papers have addressed the state of software testing technique evaluations. Juristo et al. [3] examined 25 years of empirical studies evaluating techniques in order to compile empirical results and assess the maturity level of knowledge for different testing technique families. More specifically, major relevant contributions were collected and schemes were defined for classifying them based on the family of the testing technique they evaluated and the type of evaluation they performed (laboratory study, formal analysis, laboratory replication, or field study). Each contribution was then analyzed at a depth sufficient for (1) summarizing significant implications of its empirical results and (2) categorizing it based on the two schemes mentioned above. In doing so, Juristo et al. [3] were able to summarize major findings from empirical studies in each testing technique family; providing valuable guidance for selecting testing techniques. By classifying contributions based on the type of evaluation performed, they were also able to summarize what types of empirical evaluations had been performed on each testing technique family. The method used by Juristo et al. [3] is generally referred to as a literature review. While literature reviews also compile and examine studies in a research area, they tend to differ from systematic mapping studies in study selection process, depth of study analysis, and scope:

- *Study selection process*: The process for finding and selecting relevant studies for analysis in systematic mapping studies is unambiguous and systematic. This means that sources used for searching for studies, strings used in the search, and criteria for determining whether or not a paper is relevant are all well-defined. Our systematic mapping study process is defined in greater detail in Sections 3.2 and 3.3. On the other hand, literature reviews that are not systematic literature reviews may not rigorously define their search and selection process, if at all. This marks one major difference between our study and the literature review performed by Juristo et al. [3] which does not describe a systematic method for finding important studies with testing technique evaluations.

- *Depth of study analysis*: After selecting a set of relevant papers, both systematic mapping studies and literature reviews involve some level of analysis of each paper to answer their respective research questions. In systematic mapping studies, each paper is examined only enough to map that paper into categories based on predefined classification schemes. Literature reviews tend to analyze relevant papers at a greater depth and in more detail to achieve research goals that would not be possible performing only classification. For example, Juristo et al. [3] examine relevant studies in more detail than our mapping study to be able to extract major empirical findings from each study. This allows them to summarize applicability conditions for various testing techniques based on empirical and significant evidence.

- *Scope*: Due to differences in effort required to analyze each paper in systematic mapping studies and literature reviews, the size of the research field being mapped or reviewed also tends to differ. With less analysis effort required for each relevant study, mapping studies allow the structuring of a fairly large research area. For example, our mapping study considers all testing technique evaluations across four of the most common online scientific databases.

In general, our study differs from Juristo et al. [3] in that it systematically gathers a larger set of papers in the field and categorizes them according to different classification schemes better suited for our research goals (Table 1). This approach provides assistance for answering a broader range of finer-grain questions regarding testing technique evaluations by pointing researchers to sets of actual papers belonging to more specific categories they are interested in.

Table 1 Methods for assessing the state of a research field.

Method	Study collection process	Scope	Analysis depth and goals
Systematic mapping study	Well-defined, systematic, and relatively unbiased	Potential for larger field under study due to less analysis effort per collected study.	Each collected study analyzed at a depth only sufficient for classification based on defined schemes. This depth of analysis is suitable for indexing the field and revealing the distribution of research by categories.
Systematic literature review	Well-defined, systematic, and relatively unbiased	Field under study tends to be smaller due to larger analysis effort per collected study	Each collected study analyzed in greater detail; allowing researchers to satisfy goals of identifying best practices or summarizing evidence in the field.
Literature review	Ambiguous and potentially biased	Field under study tends to be smaller due to larger analysis effort per collected study	Each collected study analyzed in greater detail; allowing researchers to satisfy goals of identifying best practices or summarizing evidence in the field.

Gonzalez et al. [4] extended the work of Juristo et al. [3] by performing a more recent examination of testing technique experiments with similar goals. The extension is similar to our research in that it utilizes a systematic mapping study to develop an understanding of the state of testing technique evaluations. Our research goals are somewhat different in that we place a particular emphasis on assisting researchers in determining how to evaluate software testing techniques in specific contexts and do not only consider experiments. For this reason this article provides a great deal of distinct information due to major differences in scope and classification schemes. In terms of scope, it includes other common evaluation methods such as case studies and does not exclude a large number of articles that report smaller experiments. It also includes articles providing guidelines or proposals regarding how testing techniques should be evaluated. In terms of classification schemes, we utilize six distinct schemes and some additional secondary categorizations of these schemes. Due to these deviations this chapter is able to answer different research questions that align more with our desire to help researchers in evaluating their testing technique.

Gonzalez et al. [4] mention two other reports from Engstrom et al. [5] and Singh et al. [6] that are systematic literature reviews of regression testing technique evaluations. Engstrom et al. [5] systematically collected papers in a way similar to our systematic mapping study to find 27 papers empirically evaluating regression test selection techniques. These papers were classified by regression test selection technique being evaluated and by the research approach used for the evaluation. Instead of completely relying on previous classification schemes for regression testing technique types, Engstrom et al. [5] used properties of testing techniques identified in the collected relevant papers to develop its own regression test selection technique classification scheme. We employ a similar technique called *keywording* for coming up with some classification schemes in our mapping study. This technique is discussed more in Section 4.1. After classifying the papers, Engstrom et al. [5] performed a full text analysis on each one to assess its empirical quality and significance. They found the empirical foundation for comparing regression test selection techniques to be fairly weak and sometimes contradictory. The other systematic literature review mentioned by Gonzalez et al. [4] and Singh et al. [6] examines regression test prioritization technique evaluations. By classifying systematically collected studies by tools and artifacts used, they identify major research gaps in empirical evaluations with certain tools and artifacts. After a detailed analysis of the empirical evidence, they are also able to summarize which technique types generally perform better than others. Our study does not include regression testing selection or prioritization techniques since we are mainly interested in the evaluation of fault-detecting software testing techniques. Jia and Harman [7] is a comprehensive systematic literature review that provides a high level overview of the mutation testing field. In it, over 390 mutation testing papers published from 1977 to 2009 were systematically collected by defining search strings and applying the strings to common online databases. Classifications on publication information and numerous other study properties were performed in order to develop a high level overview of research distributions by categories. In addition to this, papers were analyzed further in order to summarize research in the field related to mutation testing computational cost reduction techniques, equivalent mutant detection techniques, empirical evaluations in mutation testing, and barriers in mutation testing research. While we are interested in the state of mutation testing evaluations, we are interested in the evaluation of other testing techniques as well. Being a systematic mapping study, our study also places more emphasis on the classification of relevant studies than on performing a more detailed analysis for summarizing their content and significance. This approach better supports our goal of assisting researchers in locating evaluations with specific properties.

Finally, an article by Briand [8] discusses the common threats to validity of empirical studies evaluating the cost-effectiveness of software testing techniques and how they can be reduced. While this critical analysis of the field does not cite research containing threats to validity (potentially to avoid bringing attention to specific studies that did a poor job in this regard), it points out typical pitfalls in software testing empirical evaluation research. These include construct validity issues related to measuring test cost and test effectiveness, ambiguous external validity due to a lack of research across a wider variety of artifacts, and a lack of consideration of variability across test suites that satisfy the same coverage criteria. Our mapping study does not investigate each collected primary study deeply enough to confirm threats to validity that are common to certain evaluation types, but it may similarly provide some insight on the quality of current evaluations. It does so in classifying papers with case study and experiment evaluations by whether or not they possess what are considered to be exemplary case study and experiment properties; revealing the distribution of evaluations in the field with these properties. Our study additionally brings awareness to other papers in the field that provide guidelines or propose enhancements when it comes to evaluating software testing techniques.

3. Research method

An overview of the systematic mapping process is illustrated in Fig. 1. Each step of the process is described in more detail in Section 3. At a high level, we define research questions from our research goals, systematically gather a set of particles that are ideally representative of the field of interest, and then map the articles into defined categories in order to structure the field and answer our research questions.

3.1 Definition of research questions

We begin by deriving research questions from the main goals of this study. As stated in the introduction, we would like to structure the field of research evaluating software testing techniques and develop an understanding of what is state of the art by identifying and analyzing reports in the field. The following questions are derived from the goals.

RQ 1. What are the publication trends in research evaluating software testing techniques?

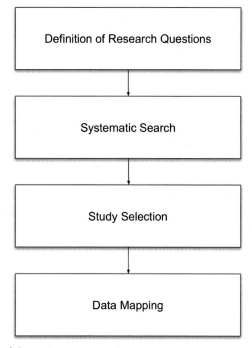

Fig. 1 Overview of the systematic mapping process.

(i) *RQ1.1*: What is the annual number of publications in the field?

(ii) *RQ1.2*: What are the main publication venues?

(iii) *RQ1.3*: What is the distribution of reports in terms of academic or industrial affiliation?

RQ 2. What is the current state of the art when it comes to evaluating software testing techniques for their effectiveness, efficiency, applicability, and scalability and where are there research gaps?

(i) *RQ2.1*: What methods have been used or proposed for evaluating software testing techniques?

(ii) *RQ2.2*: What is the distribution of methods used for evaluating software testing techniques?

(iii) *RQ2.3*: What is the distribution of dimensions being evaluated?

(iv) *RQ2.4*: What is the distribution of evaluations of white-box vs black-box testing techniques?

(v) *RQ2.5*: What can we say about the relative quality of evaluations?

(vi) *RQ2.6*: What is the distribution of reports in terms of contribution type?

(vii) *RQ2.7*: What is the distribution of effectiveness evaluations utilizing mutation analysis?

3.2 Systematic search

The next step of the mapping study process is to gather a set of papers that are potentially relevant to the field of interest. We do so by systematically defining a search string, identifying important scientific databases, and then applying the search string to the identified databases to retrieve papers.

Similar to the systematic literature review performed by Nair et al. [9], our search string was derived by first splitting up the phenomena under investigation into major terms. For each major term, keywords synonymous with the term were added using the OR operator. Keywords were heavily influenced by our research questions and research goal scope. We joined the major terms using the AND operator. The resulting search string was iteratively refined by assessing its ability to generate relevant papers from small subsets of papers and modifying keywords accordingly. We arrived at the following search string:

(evaluate OR validate OR assess) AND (effectiveness OR efficiency OR applicability OR scalability) AND ("software testing" OR "software verification" OR "black-box testing" OR "white-box testing") AND (techniques)

For scientific databases we selected some of the most common online sources: ACM, IEEE Xplore, Springer, and Wiley.

Due to our fairly broad scope and interest in the current state of the art and research gaps, we limited our search to only include reports published within the last 11 years [2007–2017]. We also excluded books from our search results since we are interested in scholarly peer-reviewed work that is more likely to be of higher quality. Only one paper was excluded due to being written in a language other than English (the language the researchers carrying out the mapping study could read). We applied our search string to each of the online databases to obtain 7426 potentially relevant papers.

3.3 Study selection

The study selection process entails removing irrelevant studies. Fig. 2 illustrates our study selection process along with the number of papers remaining after applying each step in the process.

We began by applying title and abstract exclusion. This excludes papers that are deemed irrelevant based on the content of their title and abstract. We will refer to the criteria used to assess a paper's relevance as the content criteria. Our content criteria is heavily influenced by the research goals and their scope. A paper was deemed relevant if it (1) proposed a method or guidelines for evaluating a failure-detecting software testing technique's effectiveness, efficiency, applicability, or scalability or (2) utilized a method for evaluating a failure-detecting software testing technique's effectiveness, efficiency, applicability, or scalability. As we are only interested in failure-detecting techniques, software testing techniques that do not detect failures such as test case prioritization and fault localization are not considered. This criteria included papers evaluating a developed tool, given that the tool implemented some failure-detecting software testing technique. If it could be determined that a paper did neither (1) or (2) based on its title and abstract, it was considered irrelevant and excluded from the rest of the systematic mapping process. For some papers it was unclear whether or not they satisfied the content criteria solely from their abstract and title. Text skimming was applied to such papers until the researchers could confidently assert that the paper was relevant or irrelevant. Duplicates were also removed.

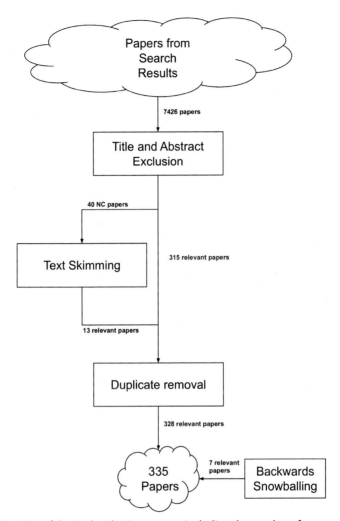

Fig. 2 Overview of the study selection process including the number of papers resulting from each step.

To reduce the threat of missing relevant papers, we applied backwards snowballing [10] to a small subset of the relevant papers. We found seven more relevant studies. In all, 335 primary studies were identified in the study selection process.

3.4 Data mapping

The final step involves mapping the relevant papers into categories based on well-defined classification schemes. The classification schemes are defined in detail along with how they were constructed in Section 4. Each relevant paper was skimmed to the extent necessary to categorize the article according to each classification scheme.

4. Classification schemes

In this section we provide the classification schemes used for the data mapping. The data facets that schemes were developed for were derived from our research questions. For example, to answer research question 1.1, "What is the annual number of publications in the field?", papers were categorized based on the year in which they were published.

The publication year, publication venue, and affiliation of the authors were extracted to answer research questions related to general publication trends. The *evaluation method, evaluation dimension, testing technique type, contribution type*, and *usage of mutation analysis* were extracted to answer more context-specific research questions. The classification schemes for these facets are discussed in more detail in Section 4.

4.1 Evaluation method

The evaluation method scheme categorizes papers based on the method they use for evaluating a software testing technique. We systematically determined evaluation method categories using *Keywording* as suggested by Petersen et al. [2]. This consisted of reading the abstracts of a subset of the collected relevant papers and generating keywords for the evaluation methods. After reading a fairly large number of abstracts, the generated keywords were clustered to form categories for methods of evaluating software testing techniques. In our case there were few unique keywords, most of which referred to fairly well-defined methods in research. Thus we relied on existing definitions to classify the four major categories:

1. *Experiment*: A paper was classified in the experiment category if it utilized an experiment to evaluate a software testing technique. This determination relied heavily on Wohlin's definition of experiments as an empirical investigation in which "different treatments are applied to or by different subjects, while keeping other variables constant, and measuring the effects on outcome variables" [11].

2. *Case study*: A paper was classified in the case study category if it utilized a case study to evaluate a software testing technique. A case study is "an empirical enquiry that draws on multiple sources of evidence to

investigate one instance (or a small number of instances) of a contemporary software engineering phenomenon within its real-life context" [12].

3. *Example*: A paper was classified in the example category if it utilized an example to evaluate a software testing technique. We define an example as a demonstration of a single technique in a small and constructed context.

4. *Analytic*: A paper was classified in the analytic category if it utilized a direct evaluation of a technique based on its clear or provable properties.

Some papers utilized multiple methods for evaluating software testing techniques, so it was possible for a single paper to be placed in multiple categories. On the other hand, a small number of papers discussed guidelines or enhancements when evaluating techniques without actually utilizing an evaluation method. For example, a paper discussing experiment subject selection is a relevant paper since it provides insight on evaluating the effectiveness of a fault-detecting software testing technique, but it does not utilize a method for evaluating software testing techniques.

4.2 Evaluation dimension

Categories for evaluation dimension for this schema were derived directly from our research scope:

1. *Effectiveness*: A paper was classified in the effectiveness category if it evaluated the ability of a software testing technique to detect failures, kill mutants, or achieve some degree of coverage.

2. *Efficiency*: A paper was classified in the efficiency category if it evaluated the performance of a software testing technique in terms of speed, memory usage, or work done.

3. *Scalability*: A paper was classified in the scalability category if it evaluated how a technique performed for larger software.

4. *Applicability*: A paper was classified in the applicability category if it evaluated the ability of the technique to be applied or generalized to multiple domains.

As with the last classification scheme, it was possible for papers to be placed into multiple categories or to not fit any of the categories.

4.3 Testing technique type

This refers to the type of testing technique a paper evaluated. The categories for this scheme were directly derived from research question 2.4, which seeks to determine the distribution of white-box and black-box testing technique evaluations.

1. *White-box:* At least one of the software testing techniques evaluated is a white-box testing technique. We classify a technique as a white-box technique using a definition from Amman and Offut [13], which states that a white-box technique derives tests from the source code internals of the system under test.

2. *Black-box:* At least one of the software testing techniques evaluated is a black-box testing technique. A black-box technique derives tests from descriptions of the system under test without access to its source code internals [13]. Evaluations of gray-box testing techniques that did not require access to the source code of the software were included in this category.

For this schema, papers could be classified as belonging to both categories if both a white-box and a black-box testing technique were evaluated. Papers were also classified as belonging to both categories if the technique type of the technique being evaluated was ambiguous and the technique was potentially applicable in both black-box and white-box contexts. Thus all papers utilizing a technique evaluation were classified as at least white-box or black-box.

4.4 Contribution type

This scheme classifies papers based on the type of contribution they make in the field. We were particularly interested in the separation of papers utilizing methods as opposed to proposing new methods or guidelines for evaluating software testing techniques. We defined the following categories:

1. *Guideline:* A paper was classified as a guideline paper if it provided guidelines for evaluating a software testing technique, proposed a method for evaluating software testing techniques, or proposed an enhancement for a method of evaluating a software testing technique.

2. *Usage:* A paper was classified as a usage paper if it utilized some method for evaluating a software testing technique for its effectiveness, efficiency, scalability, or applicability.

Papers that met both criteria were classified in both categories. Every paper was classified in at least one of the contribution type categories.

4.5 Use of mutation analysis

The mutation analysis schema below categorizes papers based on whether or not they utilize mutation analysis to evaluate the effectiveness of software testing techniques:

1. *Mutation*: A paper was classified as a mutation paper if it utilized mutation analysis and evaluated the effectiveness, efficiency, scalability, or applicability of one or more software testing techniques.
2. *Not mutation*: A paper was classified in this category if it evaluated the effectiveness, efficiency, scalability, or applicability of one or more software testing techniques and did not use mutation analysis.

As a result of this classification schema, all usage papers were categorized as either mutation or not mutation papers. Guideline papers were included in this categorization since guideline-only papers did not evaluate the effectiveness, efficiency, scalability, or applicability of a testing technique.

4.6 Evaluation quality

To answer RQ2.5 (What can we say about the relative quality of the evaluation methods?), additional data was extracted: empirical evaluations in the form of case studies and experiments. For each of these methods, we relied on proper methodology guidelines to derive data facets that would help us assess the current state of evaluations in the field in terms of quality.

Guidelines for case study methodology in the field of software engineering are discussed by Runeson in [12]. They require the definition of research questions, examination of multiple perspectives, provision of a logical link between evidence and conclusions made, and a discussion of threats to the validity of the study. From these guidelines, the following categories were created:

1. *Research Questions*: A paper was classified in this category if it clearly defined research questions to be addressed by the study.
2. *Triangulation*: This category assessed the case study's consideration of multiple perspectives. A paper utilizing a case study was classified as a triangulation paper if it collected data from multiple sources or used multiple types of data collection.
3. *Threats to Validity*: A paper was classified in this category if it seriously discussed threats to the validity of the study. A discussion was considered "serious" if it presented multiple threats and was at least a paragraph in length.

It should be noted that an evaluation framework for *empirical methods* in software testing was recently developed by Vos et al. [14]. This framework is much more detailed and focused, but due to its newness in the field it was not feasible to derive categories from it for this mapping study.

Guidelines [11] for controlled experiment methodology in the field of software engineering are used to similarly develop categories for experiment papers. Some important characteristics of exemplary experiments include a clearly stated hypothesis with hypothesis testing, justification for object/subject selection, descriptive statistics, and a discussion of threats to the validity of the experiment. From these guidelines, the following categories were created:

1. *Hypothesis testing:* A paper was classified in the *hypothesis testing* category if it clearly stated a hypothesis and performed hypothesis testing to accept or reject this hypothesis.

2. *Descriptive statistics:* A paper was classified in the *descriptive statistics* category if it utilized descriptive statistics when quantitatively analyzing results.

3. *Context justification:* This category assessed the appropriateness of objects and subjects selected in controlled experiments. To meet the *context justification* criteria, a paper's objects or subjects needed to be fairly representative of the research question context, a common benchmark, or at least justified to a degree by some discussion in the paper. Thus papers presenting objects/subjects without justification for their selection or a clear connection to research goals were not included in this category.

4. *Threats to validity:* A paper was classified in this category if it seriously discussed threats to the validity of the experiment. A discussion was considered "serious" if it presented multiple threats and was at least a paragraph in length.

5. Evaluating software testing techniques: A map of the field

We present a map of the field of research evaluating software testing techniques. Three hundred and sixty-five relevant papers were systematically collected and mapped according to the classification schemes defined above; providing a large-scale overview of publication trends, research gaps, and the state of the art when it comes to evaluating software testing techniques.

5.1 Publication trends

5.1.1 Annual activity level

Fig. 3 illustrates the level of activity in the field over the last 11 years. The annual number of relevant papers increased significantly from 2009 to 2011 before fluctuating over the 7 remaining years of the mapping. As shown by the black line of best fit, the annual number of published papers in the field has grown a good amount overall. This suggests an increased interest in research evaluating software testing techniques.

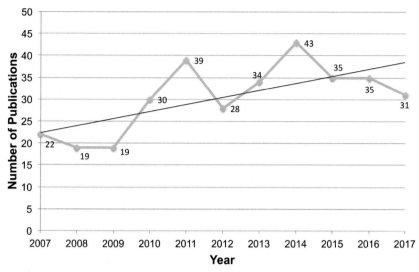

Fig. 3 Annual number of publications.

Table 2 Main publication venues.

Publication venue	#	%
Software Testing, Verification and Reliability	33	9.85
International Symposium on Software Testing and Analysis	24	7.16
International Conference on Automated Software Engineering	17	5.07
International Conference on Software Engineering	15	4.48
International Conference on Software Testing	15	4.48
International Symposium on Foundations of Software Engineering	14	4.18
Empirical Software Engineering	13	3.88
International Conference on Software Testing, Verification and Validation	13	3.88

5.1.2 Main publication venues

Not surprisingly given our fairly broad research scope, the relevant papers collected spanned 120 unique publication venues. While many of these venues only published one relevant paper, there were some venues responsible for publishing a significant number of contributions in the field. Table 2 lists the venues that published the most relevant papers along with how many they published. By a significant margin, the journal of Software Testing,

Verification and Reliability was the most active publication venue with 33 relevant papers published over the last 11 years. The International Symposium on Software Testing and Analysis was the next largest contributor with 24 relevant papers. Six other venues listed in Table 2 had 10–20 relevant publications. The remaining venues had less than 10 relevant publications, with 79 venues having only 1 relevant publication.

5.1.3 Industry vs academia

Fig. 4 shows the relative contributions of industry and academia based on author affiliation. Similar to most fields of research, a large majority of contributions are made by academia. Two hundred and ninety-one papers (about 87%) had exclusively authors affiliated with academic institutions. Thirty papers (about 9%) had both authors affiliated with academic institutions and authors affiliated with industry. Only 14 papers (about 4%) had exclusively authors affiliated with industry.

5.2 Context-specific mappings

5.2.1 Evaluation method

We developed 4 major categories for methods of evaluation: experiments, case studies, examples, and analytic evaluations. Fig. 5 shows the number of papers that utilized each evaluation method. Percentages shown are of the total number of evaluation instances as opposed to the total number of primary studies. As mentioned earlier, case studies and controlled experiments were by far the most common methods. Experiments in particular were utilized very frequently for evaluating software testing techniques. Of the

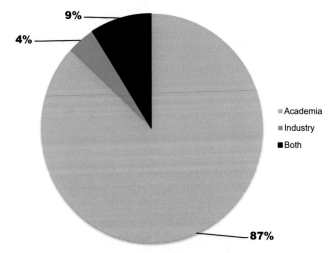

Fig. 4 Percentage of contributions from industry and academia.

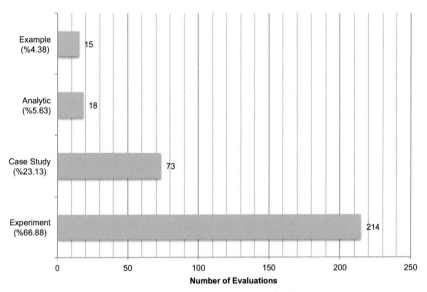

Fig. 5 Distribution of primary study evaluations by method.

320 instances of testing technique evaluations, 214 of them (66.88%) were controlled experiments. Seventy-three of them (23.13%) were case studies. Only 18 were analytic evaluations and only 15 fell into the example category. From this one data facet it seems that performing controlled experiments is the state of the art when it comes to evaluating software testing techniques. Exploring the relation between evaluation methods and other data facets provides more insight on how the state of the art changes with the dimension and type of testing technique evaluated.

5.2.2 Evaluation dimension

We also categorized papers based on the dimension they evaluated (effectiveness, efficiency, applicability, and scalability). Fig. 6 shows the number of evaluations performed for each dimension. Percentages shown are of the total number of dimension evaluations. Note that there are more dimension counts than the number of relevant papers collected since some papers evaluated more than one dimension of a software testing technique.

Of the 425 total dimension evaluations, more than half of them (55.06%) evaluated effectiveness; suggesting that researchers are the most interested in evaluating techniques based on their ability to detect failures, kill mutants, or achieve some degree of coverage. This makes sense given the main purpose of testing techniques to reduce the risk of using software by detecting failures.

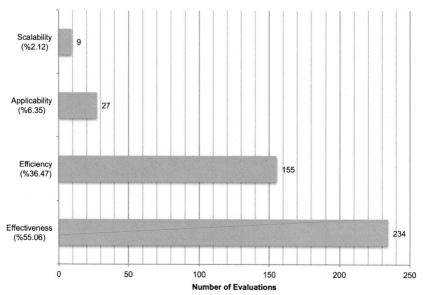

Fig. 6 Number of evaluations by dimension.

Another large portion of the total dimension evaluations (36.47%) assessed the efficiency of a technique. The remaining 8.47% is split between applicability and scalability evaluations at 6.35% and 2.12%, respectively.

5.2.3 Testing technique type

Fig. 7 illustrates the distribution of testing technique types that were evaluated. We see that research evaluating software testing techniques is quite evenly split between white-box and black-box testing techniques. About (46.71%) of papers with evaluations are focused on white-box testing techniques, 49.01% are focused on black-box testing techniques, and the remaining 4.28% evaluated both of these testing technique types. While there are a good portion of papers dealing with the evaluation of black-box testing techniques, we found that a large chunk of these evaluations are of the same few techniques. Upon further investigation, about 30% of the black-box evaluations were of random testing or combinatorial testing techniques.

5.2.4 Contribution type

This scheme classified papers based on whether they evaluated a software testing technique or proposed some method or insights regarding how software testing techniques should be evaluated. As Fig. 8 illustrates, the majority of papers were usage papers that utilized some method for evaluating

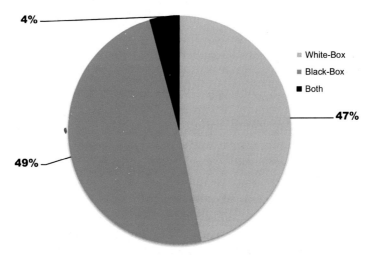

Fig. 7 Percentage of white-box and black-box evaluations.

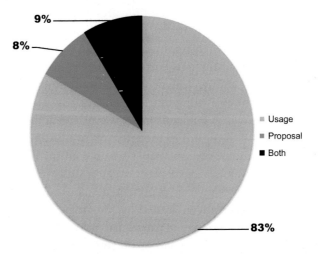

Fig. 8 Contribution type distribution.

software testing techniques. On the other hand relatively very few papers discussed how techniques should be evaluated or proposed a new methodology for doing so.

Despite the lower number of proposal papers, we believe this type of contribution to the field is important for evolving and enhancing our ability to assess software testing techniques. As such, a secondary classification was performed on these papers to develop an understanding of the types of insights and proposals that exist for evaluating software testing techniques and to be able to point researchers toward higher level guidelines in areas they are interested in. Our hope is that bringing awareness to these papers will allow researchers to make higher quality evaluations of testing techniques as well as motivate more research of this contribution type. The following section describes the secondary classification and presents the results.

5.2.5 Guideline paper classification

Due to a lack of existing knowledge regarding the types of guideline papers we would find, *keywording* [2] was again used to develop categories for the types of guideline papers after examining each one in more detail. Doing so resulted in the following classification schema:

1. *Program artifact*: Program artifact papers provide guidelines or insight for program artifacts under test when empirically evaluating software testing techniques. These include papers discussing the importance of considering fault types in effectiveness evaluations, advocating for common benchmark artifacts, and the state of the art in software fault injection.

2. *Evaluation metric*: Evaluation metric papers provide guidelines or insight for choosing a metric when empirically evaluating software testing techniques. These include empirical correlations between evaluation metrics and fault-detecting ability, analytic effectiveness bounds, and proposals of novel criteria for evaluating test suite quality.

3. *Human subject selection*: A guideline paper was placed in this category if it provided insight with regards to the selection of human subjects for an empirical evaluation. Only one paper was placed in this category for exploring the impact of subject experience on study results.

4. *Methodology*: Methodology papers presented empirical study methodology guidelines not already addressed by the artifact or human subject selection categories. Examples of papers mapped to this category include the proposal of a unified framework and an outline of proper methodology when conducting empirical evaluations in software testing.

5. *Mutation analysis (code)*: Code mutation analysis papers presented an innovation or guideline to mutation analysis of test suites at the source code level. In some form they provided suggestions for how mutation analysis should be performed. Most of these papers discuss efficiency improvements as this is a well-known limitation of mutation analysis techniques. We separate mutation testing at the code level from mutation testing at the model level due to the large number of mutation analysis guideline papers and significant differences in guidelines between the two.

6. *Mutation analysis (model)*: Model mutation analysis papers presented an innovation or guideline to mutation analysis of test suites at the model level.

A full text skimming was applied to each of the proposal papers as they were categorized using the above schema. Table 3 presents the results of the secondary categorization; mapping each category to a set of proposal papers belonging to it. Furthermore, a short summary is provided with each of the proposal papers to make it easier for researchers to locate papers relevant to their interests.

5.2.6 Use of mutation analysis

Fig. 9 illustrates the portion of evaluation papers classified as mutation papers. Of all 217 papers evaluating the effectiveness of a testing technique using a case study or experiment, a large portion of them (28%) utilized mutation analysis. Furthermore, mutation analysis seems to be becoming more popular over time. Fig. 10 shows the proportion of effectiveness evaluations that utilize mutation analysis each year. One of the main limitations of mutation testing and analysis has been its high computational cost. It makes sense that mutation analysis has become more popular as more cost reduction strategies are developed and refined.

5.2.7 Evaluation quality

Tables 4 and 5 present the results of extracting evaluation quality data facets from experiments and case studies, respectively.

Very few experiments (18%) formally stated a hypothesis and performed hypothesis testing. On the other hand, a majority of experiments utilized descriptive statistics. We see that close to half of experiments meet the justified context criteria and provide a serious discussion of threats to validity. A smaller percentage of case studies provided threats to validity. 57% of case studies implemented some form of data triangulation while few (27%) clearly stated research objectives.

5.2.8 Distribution of evaluation methods over time

Fig. 11 shows the distribution of evaluation methods over time. Experiments were the most common method of evaluating testing techniques every year. The number of case study and experiment evaluations grew considerably from 2007 to 2014; growing by 366.67% and 236.36%, respectively. The number of experiments and case studies remained fairly high in the last 3 years of the study. Both the number of examples and analytic evaluations remained low throughout the study with minor variation.

Table 3 Guideline papers by category.

Guideline category	Papers
Program artifact	*Assessing dependability with software fault injection: A survey* [15] presents an overview of the state of the art in software fault injection and insight on which approaches to apply in different contexts.
	BegBunch: benchmarking for C bug detection tools [16] presents two benchmark programs in the C language with the hopes of providing a "common ground" for empirical comparisons of different fault-detecting techniques.
	On the improvement of a fault classification scheme with implications for white-box testing [17] presents improvements for a fault classification scheme with the notion that testing techniques are better at finding certain types of faults than others. This paper is included in the artifact selection category since considering the nature of faults in artifacts used in empirical studies may enhance our understanding of the effectiveness of software testing techniques.
	On the number and nature of faults found by random testing [18] an evaluation of the nature of faults that are discovered by random testing. Also provides a fault classification scheme and evidence that the nature of faults should also be considered when comparing testing techniques.
Evaluation metric	*An upper bound on software testing effectiveness* [19] provides an analytic upper bound on the effectiveness of software testing techniques that rely on failure patterns.
	Assertions are strongly correlated with test suite effectiveness [20] empirically evaluates the relationship between the fault-detection ability of a test suite and its assertions.
	Comparing non-adequate test suites using coverage criteria [21] an empirical evaluation in an attempt to answer which criteria should be used to evaluate test suites, particularly when test suites are nonadequate.
	Evaluating test suite effectiveness and assessing student code via constraint logic programming [22] suggests the evaluation of test suites by comparing their effectiveness with a suite automatically generated by Constraint Logic Programming
	Information gain of black-box testing [23] introduces a novel coverage criteria for assessing black-box tests based on information gain from test cases.
	On use of coverage metrics in assessing effectiveness of combinatorial test designs [24] investigates the use of certain coverage metrics when evaluating combinatorial testing strategies. Due to somewhat variable coverage across contexts for a given strategy, suggests some measure of variability should be included when assessing the effectiveness of strategies using these metrics.

Table 3 Guideline papers by category.—cont'd

Guideline category	Papers
	PBCOV: a property-based coverage criterion [25] proposes a new property-based criterion for assessing the adequacy of test suites.
	State coverage: A structural test adequacy criterion for behavior checking [26] proposes state coverage, a new structural criterion for assessing the adequacy of test suites.
	Structural testing criteria for message-passing parallel programs [27] introduces a novel structural testing criteria specifically for message-passing parallel programs. Additionally presents a tool that implements the new criteria along with results from applying it.
	The risks of coverage-directed test case generation [28] an empirical evaluation of structural coverage criteria. Among other things, concludes that traditional structural coverage criteria by itself may be a poor indicator of a test suite's fault-detection capabilities and that Observable MC/DC may be a promising alternative.
	Selecting V&V technology combinations: how to pick a winner? [29] proposes a systematic method for evaluating verification and validation technique combinations.
	Toward a deeper understanding of test coverage [30] suggests coverage criteria should be calculated at different testing levels instead of for the test suite as a whole.
	Web application fault classification—an exploratory study [31] introduces a web application fault classification schema based on the exploration of two large, real-world web systems.
Human Subject	[32] *The Impact of Students' Skills and Experiences on Empirical Results: A Controlled Experiment with Undergraduate and Graduate Students* A controlled experiment investigating how the experience of human subjects in empirical studies evaluating effectiveness and efficiency can impact results.
Methodology	[33] *Toward a Semantic Knowledge Base on Threats to Validity and Control Actions in Controlled Experiments* Proposes a knowledge base of threats to validity to assist researchers in mitigating threats when planning experiments.
	[34] *The role of replications in Empirical Software Engineering* Identifies types of empirical study replications, discusses the purpose of each type, and gives guidelines for providing sufficient information about reported empirical studies to better enable study replication.

Continued

Table 3 Guideline papers by category.—cont'd
Guideline
category   **Papers**

	[8] *A Critical Analysis of Empirical Research in Software Testing* Provides a critical analysis of empirical research in software testing and discusses common threats that arise when determining cost-effectiveness of a technique via empirical research.
	[35] *Toward Reporting Guidelines for Experimental Replications: A Proposal* Suggests publishing guidelines for experiment replications in order to "increase the value of experimental replications".
	[36] *Empirical Evaluation of Software Testing Techniques in an Open Source Fashion* Presents and advocates for a unified framework for testing technique evaluations to ease study replication and improve reproducibility of results.
	[14] *A Methodological Framework for Evaluating Software Testing Techniques and Tools* Defines a general methodological evaluation framework for case studies in software testing.
Mutation analysis (c)	*A generic approach to run mutation analysis* [37] introduces a generic approach for mutation analysis that is not restricted to particular execution environments.
	An approach for experimentally evaluating effectiveness and efficiency of coverage criteria for software testing [38] provides guidelines and a demonstration of how to evaluate the effectiveness and efficiency of coverage criteria utilizing mutation analysis.
	Do redundant mutants affect the effectiveness and efficiency of mutation analysis? [39] empirically demonstrates efficiency and effectiveness improvement gains from removing redundant mutants in mutation analysis.
	Efficient mutation testing of multithreaded code [40] introduces a general framework for efficient exploration that can reduce the time for mutation testing of multithreaded code.
	Extended firm mutation testing: a cost reduction technique for mutation testing [41] discussion of various mutation cost reduction techniques and a proposal for a new execution based cost reduction technique.
	Faster mutation testing inspired by test prioritization and reduction [42] proposes a mutation testing cost reduction technique that prioritizes tests to more quickly determine which mutants were killed.
	Measuring effectiveness of mutant sets [43] empirical investigation and guidelines regarding how mutant sets should be evaluated.

Table 3 Guideline papers by category.—cont'd

Guideline category	Papers

Mutants generation for testing Lustre programs [44] presents a mutation generator for Lustre programs that employs mutation cost reduction techniques.

Mutation testing in practice using ruby [45] presents mutation operators for Ruby and guidelines for mutation testing based on experience from an industrial Ruby project.

Mutation testing strategies using mutant classification [46] proposes mutant classification strategies to assist in isolating equivalent mutants along with an experimental evaluation of the technique.

Mutation testing techniques: a comparative study [47] an empirical comparison of four mutation testing techniques (operators at class level, operators at method level, all operators, and random sampling)

The major mutation framework: efficient and scalable mutation analysis for Java [48] introduces a JUnit mutation analysis and fault seeding framework with claims of scalability and efficiency.

The use of mutation in testing experiments and its sensitivity to external threats [49] brings to light important external threats to consider when utilizing mutation testing in experiments. These threats may be caused by test suite size, selected mutation operators, and programming languages.

Using evolutionary computation to improve mutation testing [50] introduces a mutation testing cost reduction technique that utilizes a genetic algorithm to produce a reduced set of mutants.

An empirical evaluation of the first and second order mutation testing strategies [51] provides an evaluation of the cost and effectiveness of different mutation testing strategies.

Decreasing the cost of mutation testing with second-order mutants [52] proposes a cost reduction technique for mutation testing/analysis that combines mutants from an original set to obtain a new set of mutants. Additionally performs an empirical evaluation of a test suite created from these combined mutants.

Efficient JavaScript mutation testing [53] proposes mutation operators specific to web applications and a mutation cost reduction technique.

Efficient mutation analysis by propagating and partitioning infected execution states [54] significant efficiency gains in mutation analysis using state infection conditions. The approach is also implemented and empirically evaluated on open source programs.

Continued

Table 3 Guideline papers by category.—cont'd

Guideline category	Papers
	Evaluating mutation testing alternatives: a collateral experiment [55] proposes second-order mutation strategies and provides experimental results suggesting the strategies lead to significant cost reductions without considerably reducing test effectiveness.
	Exploring hybrid approach for mutant reduction in software testing [56] introduces a hybrid mutation testing cost reduction technique.
	JDAMA: Java database application mutation analyser [57] introduces a mutation analyzer useful for evaluating testing techniques applied to java database applications.
	Mutation operators for Simulink models [58] proposes a set of mutation operators for Simulink models and provides a procedure for mutation testing of Simulink models.
	Mutation operators for the Atlas Transformation Language [59] presents mutation operators for the Atlas Transformation Language and evaluates their effectiveness in an empirical study.
	Parallel mutation testing [60] suggests enhancing the efficiency of mutation testing by utilizing parallel execution.
	Reducing mutation costs through uncovered mutants [61] presents a mutation cost reduction technique that leverages the analysis of covered mutants to reduce the number of executions required.
	Selective mutation testing for concurrent code [62] "explores selective mutation techniques for concurrent mutation operators" and provides an empirical study evaluating these techniques.
	Speeding-up mutation testing via data compression and state infection [63] speeds up mutation testing by filtering out executions using state infection information and grouping mutants with Formal Concept Analysis.
	Statistical investigation on class mutation operators [64] provides statistical information regarding the number of mutants generated, the distribution of mutants generated, and the effectiveness of applying class mutation operators to 866 open source classes.
	Topsy-Turvy: a smarter and faster parallelization of mutation analysis [65] presents a new parallelization technique for mutation analysis.
	Using conditional mutation to increase the efficiency of mutation analysis [66] introduces a new efficiency optimization when performing mutation analysis called conditional mutation.
	X-MuT: a tool for the generation of XSLT mutants [67] introduces mutation operators for the XLST language along with their implementation in a tool and an evaluation of its effectiveness.

Table 3 Guideline papers by category.—cont'd

Guideline category	Papers
Mutation analysis (m)	*A variability perspective of mutation analysis* [68] introduces method for modeling mutation operators as a feature diagram for better and faster mutation analysis.
	Featured model-based mutation analysis [69] proposes an optimization for model-based mutation analysis using a modeling framework. Performance evaluations of the proposed technique are carried out and compared to other optimizations.

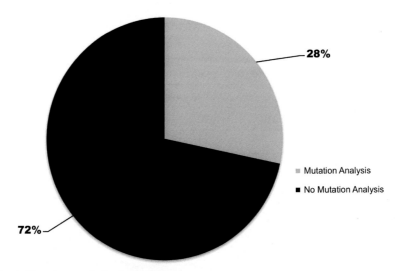

Fig. 9 Mutation analysis distribution.

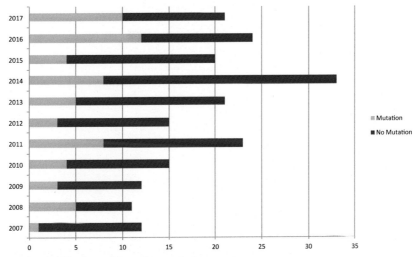

Fig. 10 Distribution of mutation analysis over time.

Table 4 The number and percent of experiments that satisfy each of the experiment evaluation quality criteria.

Experiment evaluation quality

Category	# of experiments	%
Hypothesis testing	39	18.22
Context justification	98	45.79
Descriptive statistics	160	74.77
Threats to validity	100	46.73

Table 5 Number and percent of case studies that satisfy each of the case study evaluation quality criteria.

Case study evaluation quality

Category	# of case studies	%
Research questions	20	27.40
Triangulation	42	57.53
Threats to validity	27	36.99

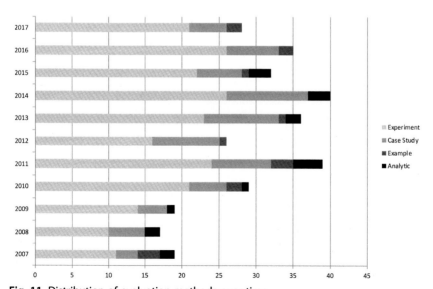

Fig. 11 Distribution of evaluation methods over time.

5.2.9 Relation of evaluation method and dimension

Table 6 gives the number of relevant papers by evaluation method and evaluation dimension. Fig. 12 illustrates their distribution. Note that the total number of papers is greater than 335 since a paper could utilize multiple evaluation methods or evaluate multiple dimensions. Given that experiments were the most common evaluation method and effectiveness was the most common evaluation dimension, it is not surprising that experiments evaluating the effectiveness of a technique are the most common here. Experiments evaluating the effectiveness and efficiency of testing techniques make up over half of the total testing technique evaluations. We see that relatively very few experiments evaluated the scalability or applicability of testing techniques. A large number of case studies also evaluate the effectiveness and efficiency of software testing techniques. Despite the much lower

Table 6 Distribution of papers by evaluation method and evaluation dimension.

	Experiment	Case study	Example	Analytic
Effectiveness	161	57	8	8
Efficiency	123	30	0	7
Scalability	5	3	0	1
Applicability	2	14	8	4

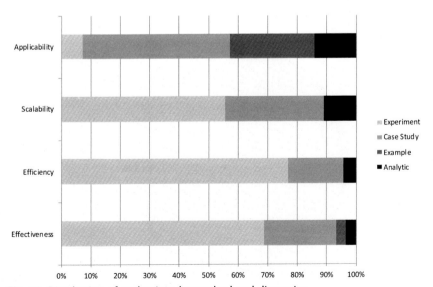

Fig. 12 Distribution of evaluations by method and dimension.

number of applicability evaluations in general (6.5% of all evaluations), 13.46% of case studies evaluated applicability. Furthermore, 50% of applicability evaluations were case studies compared to 7.14% that were experiments. Very few scalability evaluations are performed in general, but case studies and experiments make up 88.89% of them. Examples were evenly used to assess the effectiveness and applicability of techniques. No examples were used to investigate efficiency or scalability. Examples also make up a large amount of applicability evaluations (28.57%). We see that analytic evaluations assessed effectiveness and efficiency the most, but only assess scalability once.

5.2.10 Relation of mutation analysis, evaluation method, and technique type

Figs. 13 and 14 show the distribution of effectiveness papers utilizing mutation analysis in experiments and case studies. The distribution is surprisingly similar for experiments and case studies, differing only by about one percent of papers.

A somewhat greater difference can be observed when comparing the distributions of mutation analysis papers by testing technique type. Figs. 15 and 16 illustrate this difference. 33% of black–box effectiveness evaluations utilized mutation analysis. On the other hand, mutation analysis was surprisingly a bit less popular in white–box effectiveness evaluations; being used in about (25%) of these papers.

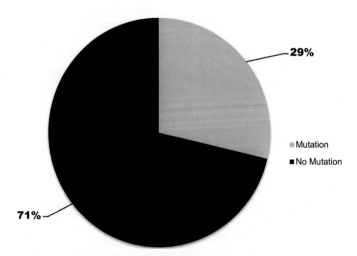

Fig. 13 Distribution of mutation analysis experiment papers.

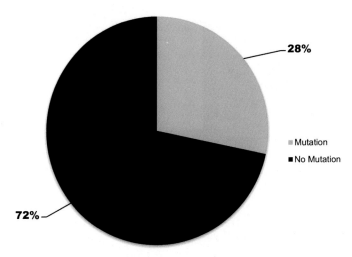

Fig. 14 Distribution of mutation analysis case study papers.

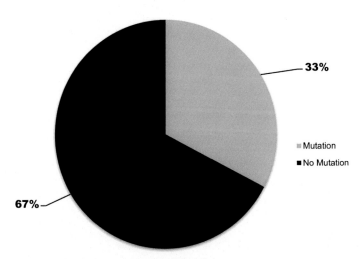

Fig. 15 Distribution of mutation analysis black-box papers.

5.2.11 Relation of author affiliation, evaluation method, and evaluation dimension

Fig. 17 shows the relation between author affiliation, evaluation method, and dimension of evaluation. We see that industry has the most involvement with experiments assessing effectiveness and efficiency and with case studies assessing effectiveness, efficiency, and applicability. Industry has little affiliation with other evaluation methods or dimensions of evaluation.

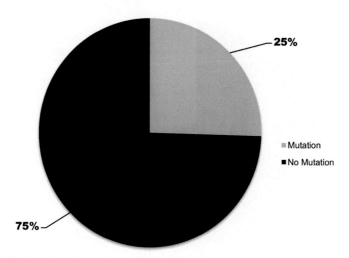

Fig. 16 Distribution of mutation analysis white-box papers.

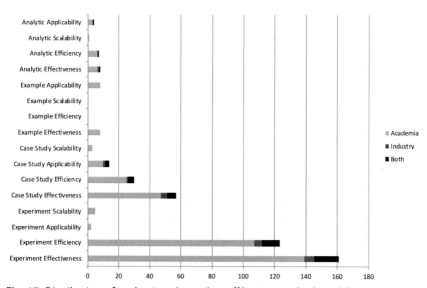

Fig. 17 Distribution of evaluations by author affiliation, method, and dimension.

5.2.12 Relation of technique type, evaluation method, and evaluation dimension

Fig. 18 shows the relation between technique type, evaluation method, and evaluation dimension. We see that for most combinations of technique type and evaluation dimension, experiments are the most common method of evaluation followed by case studies. Of notable exception are applicability

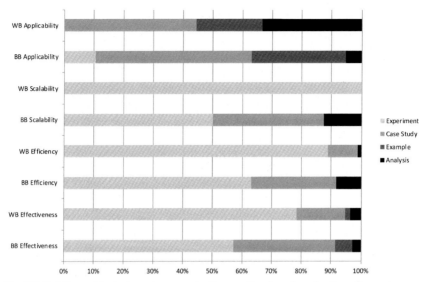

Fig. 18 Relation of technique type, evaluation method, and evaluation dimension.

evaluations of both white-box and black-box testing techniques. In these applicability evaluations, case studies become the most common evaluation method, making up 52.63% of black-box evaluations and 44.44% of all white-box evaluations. 69% of all case studies evaluating applicability were evaluations of black-box testing techniques.

More interesting are the differences between some of the evaluation method distributions with the same evaluation dimension. For instance, white-box scalability evaluations found in this study exclusively use experiments while about 50% of black-box scalability evaluations consist of case studies and analytic evaluations. Analytic evaluations also made up a greater amount of white-box applicability evaluations than they did black-box applicability evaluations. We find that across the board case study evaluations are a good amount more common when evaluating black-box testing techniques.

5.3 Papers by category

Probably the largest contribution of this paper is a map from our classifications to sets of specific papers belonging to them. We hope such a map will allow researchers to easily locate papers evaluating software testing techniques with certain characteristics. In particular, researchers looking to evaluate a particular technique can develop an understanding of how they should do so by utilizing the map to find the state of the art for similar technique evaluations.

Each combination of technique type, evaluation dimension, evaluation method, and mutation affiliation is mapped to a set of papers along with the set's cardinality in Table 7. Due to the large number of papers, each paper is presented using its citation number. Due to the large number of category combinations (64), the table utilizes a unique context identifier as a key assigned to each subset of evaluation method combinations. A complementary decision tree (Fig. 19) is provided for quickly obtaining a context identifier based on paper characteristics, and thus for quickly finding a table entry of interest since context identifiers are sorted alphabetically. The internal nodes of the tree represent classification schemes, with branches to children representing each classification in the scheme. The leaves of the tree are the context identifiers for entries in Table 7. Thus context identifiers are obtained from the tree by following a path from its root to a leaf based on classification categories of interest. A more in-depth demonstration utilizing the tree and table is presented in a case study in Section 7.

Table 7 Papers belonging to each category combination.

Context ID	Evaluation method	Count	Articles
A	Experiment	30	[70], [71], [72], [73], [55], [47], [74], [75], [76], [77], [78], [79], [62], [80], [42], [81], [82], [83], [84], [85], [38], [86], [87], [88], [89], [90], [91], [92], [93], [94]
	Case study	6	[53], [45], [95], [96], [97], [46]
	Analytic	0	
	Example	0	
B	Experiment	24	[70], [98], [60], [72], [64], [63], [55], [56], [47], [75], [54], [48], [77], [78], [79], [66], [42], [82], [85], [38], [86], [99], [90], [93]
	Case study	4	[53], [65], [96], [88]
	Analytic	0	
	Example	0	
C	Experiment	0	
	Case study	2	[45], [96]
	Analytic	1	[57]
	Example	0	

Table 7 Papers belonging to each category combination.—cont'd

Context ID	Evaluation method	Count	Articles
D	Experiment	0	
	Case study	0	
	Analytic	0	
	Example	0	
E	Experiment	71	[100], [101], [102], [103], [104], [105], [106], [52], [107], [108], [109], [110], [111], [112], [113], [114], [115], [116], [117], [118], [119], [120], [121], [122], [123], [124], [125], [16], [126], [127], [128], [129], [130], [131], [132], [133], [134], [83], [135], [136], [137], [138], [139], [140], [141], [142], [143], [144], [145], [146], [147], [148], [149], [150], [151], [152], [153], [154], [155], [156], [157], [158], [159], [160], [161], [162], [163], [164], [51], [165], [166]
	Case study	15	[167], [168], [169], [170], [22], [171], [172], [173], [174], [175], [176], [177], [178], [179], [180]
	Analytic	2	[181], [182]
	Example	5	[101], [183], [184], [185], [186]

Context ID	Evaluation method	Count	Articles
F	Experiment	54	[40], [61], [102], [103], [187], [52], [107], [108], [188], [115], [117], [189], [118], [119], [190], [191], [120], [181], [54], [48], [192], [16], [126], [127], [193], [194], [195], [129], [131], [196], [135], [136], [137], [138], [197], [198], [140], [199], [147], [200], [151], [153], [201], [202], [203], [155], [204], [157], [158], [161], [205], [164], [51], [166]
	Case study	6	[168], [65], [173], [178], [206], [180]
	Analytic	0	
	Example	1	[207]
G	Experiment	0	
	Case study	2	[176], [208]
	Analytic	1	[209]
	Example	3	[210], [182], [211]

Continued

Table 7 Papers belonging to each category combination.—cont'd

Context ID	Evaluation method	Count	Articles
H	Experiment	1	[143]
	Case study	0	
	Analytic	0	
	Example	0	
I	Experiment	22	[212], [213], [214], [215], [216], [217], [218], [219], [220], [221], [222], [223], [77], [224], [21], [225], [226], [227], [89], [93], [228], [229]
	Case study	13	[230], [231], [59], [58], [232], [97], [233], [234], [235], [236], [237], [238], [239]
	Analytic	1	[240]
	Example	1	[224]
J	Experiment	10	[213], [98], [216], [241], [221], [77], [224], [69], [242], [93]
	Case study	7	[230], [231], [233], [234], [235], [237], [243]
	Analytic	0	
	Example	1	[224]
Context ID	Evaluation method	Count	Articles
K	Experiment	0	
	Case study	0	
	Analytic	1	[67]
	Example	0	
L	Experiment	0	
	Case study	0	
	Analytic	0	
	Example	0	

Table 7 Papers belonging to each category combination.—cont'd

Context ID	Evaluation method	Count	Articles
M	Experiment	46	[100], [244], [245], [246], [247], [248], [18], [249], [250], [251], [24], [252], [253], [116], [254], [255], [256], [257], [258], [259], [260], [261], [262], [263], [264], [265], [266], [267], [268], [269], [145], [147], [270], [149], [150], [271], [272], [273], [274], [275], [276], [277], [278], [279], [280], [281]
	Case study	25	[282], [283], [284], [285], [286], [287], [288], [289], [290], [291], [292], [265], [293], [294], [295], [296], [297], [298], [299], [300], [301], [302], [303], [180], [304]
	Analytic	5	[305], [306], [307], [308], [309]
	Example	3	[246], [310], [184]
N	Experiment	39	[245], [246], [311], [312], [248], [313], [314], [315], [316], [249], [317], [252], [318], [319], [254], [320], [256], [321], [257], [258], [322], [262], [323], [266], [267], [269], [147], [270], [324], [271], [272], [325], [326], [327], [276], [277], [278], [280], [281]
	Case study	15	[328], [329], [286], [287], [288], [291], [292], [265], [296], [330], [331], [299], [332], [303], [180]
	Analytic	0	
	Example	5	[246], [314], [252], [333], [332]
O	Experiment	2	[334], [335]
	Case study	10	[336], [329], [285], [337], [338], [339], [340], [331], [335], [301]
	Analytic	5	[341], [342], [343], [344], [306]
	Example	1	[345]
P	Experiment	4	[245], [317], [261], [326]
	Case study	3	[328], [340], [297]
	Analytic	0	
	Example	1	[319]

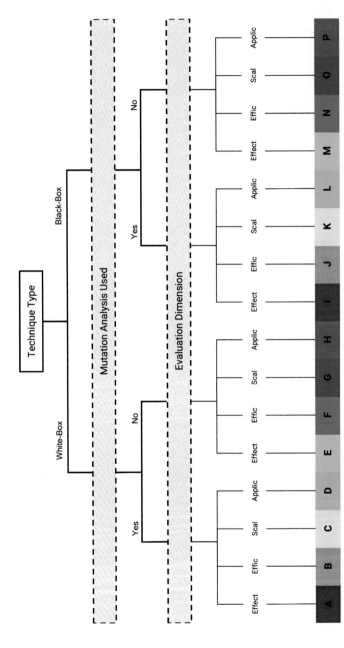

Fig. 19 Decision tree for quickly locating entries in Table 7.

6. Discussion

Our map of the field reveals that interest in research evaluating software testing techniques has grown significantly since 2007. Despite the broad scope of the field, we see that this interest does manifest itself in a few publication venues with a much higher relative concentration of relevant papers. Contributions in the field come almost entirely from academia with only a small percentage of papers written by authors affiliated with industry. Even though industrial contributions are relatively few, the distribution of evaluation methods and dimensions are somewhat different in this set of papers. A large portion of case studies examining the applicability of testing techniques from authors in industry suggests that industry can provide a valuable niche in that area.

Our study also reveals there is a good amount of research evaluating both white-box and black-box testing techniques, with about half of evaluations being of each technique type. We found that black-box technique evaluations focused largely on combinatorial and random testing techniques; leaving a relative shortage of research evaluating other black-box testing techniques. For the most part, the distribution of evaluation methods and evaluation dimensions in black-box evaluations is similar to that of white-box evaluations. That said, black-box evaluations more often utilize case studies and analytic evaluations when assessing techniques.

In general, evaluations of software testing techniques are overwhelmingly empirical studies in the form of experiments and case studies with a large focus on evaluating effectiveness and efficiency. On the other hand, there are gaps in research evaluating scalability and applicability. Based on the distribution of the dimensions of these evaluations, we can provide insight on what is the state of the art when it comes to evaluating software testing techniques:

1. For researchers looking to evaluate the effectiveness of their testing technique experiments were by far the most common methodology for doing so. Despite being the most common method of evaluation, a majority of experiments looking at the effectiveness of techniques neglected to provide a hypothesis with hypothesis testing. Only about half of experiments met the justified context criteria or provided a serious discussion of threats to validity. The second most common method for evaluating effectiveness was case studies. These were often used when research goals had to do with evaluating the technique in an industrial

context unsuitable for the level of control required for an experiment. The case studies did a poor job of meeting the case study quality criteria described in Section 4.5. Only a few papers used examples or analytic methods to demonstrate the effectiveness of their technique. In short, experiments are mostly used for evaluating the effectiveness of testing techniques when possible and experiments are so far relatively weak according to proper experiment methodology laid out by Wohlin et al. [11].

2. The state of the art is fairly similar when it comes to evaluating the efficiency of testing techniques. Experiments were again by far the most common methodology for doing so. Many of these experiments also neglected to provide hypothesis testing or discuss threats to their validity; something that can be improved upon in this field. Case studies were the second most common method used and were of similar quality to those evaluating effectiveness. A few analytic evaluations and no examples were used to assess efficiency.

3. For researchers looking to evaluate the applicability of their testing technique, case studies were the most used by a significant margin. These case studies did a better job of utilizing data triangulation and clearly defining research questions. Still, very few provided a serious discussion of threats to validity. Examples were the next most common method used for assessing applicability. These assessments tended to be simple demonstrations of how a technique could be applied in different contexts as opposed to a more rigorous empirical evaluation. Despite being the most common evaluation method, experiments evaluated the applicability of testing techniques the least. In short, case studies should be used in most cases to assess the applicability of testing techniques, with examples being used for simpler demonstrations of applicability.

4. Finally, for researchers looking to evaluate the scalability of their testing techniques, case studies and experiments were the most common methods for doing so. Even though only nine scalability evaluations were collected in this mapping study, almost all of them utilized case studies or experiments. As mentioned earlier, Scalability was a dimension in which the distribution of evaluation methods changed drastically with testing technique type. We see that the scalability of white-box techniques is only evaluated using experiments while the scalability of black-box techniques largely utilizes case studies. Thus, researchers looking to follow the state of the art when evaluating the scalability of their testing technique should consider the testing technique type when deciding between experiments and case studies.

In terms of contribution type, most of the collected papers performed an evaluation of some software testing technique. There were relatively very few papers actually discussing how techniques should be evaluated or proposing a new methodology for doing so. That said, a few important papers with the latter contribution type were presented in Section 5.2.4. These papers suggest that convergence in empirical study methodology and more careful analysis and characterization of objects to which treatments are applied will significantly improve reproducibility and the efficacy of claims made in evaluating software testing techniques.

7. Case study

To demonstrate how the results of this mapping study can be used by researchers looking to evaluate a particular testing technique, we present a small case study based on the case of our peers who are interested in evaluating the effectiveness of a novel black-box testing technique. We first introduce the case in more detail. Then we step through various sections of the results; discussing how each section helps us develop an understanding of how the novel black-box testing technique developed by our peers should be evaluated.

7.1 The case

One of the motivating examples for this mapping study came from our peers who developed a novel black-box testing technique. As with many researchers who have developed a novel testing technique, a greater understanding of how to evaluate their particular technique was desired. How have other papers evaluated similar testing techniques? Are there any best practices or guidelines to be aware of? Furthermore, the case of our peers presented a particular challenge when evaluating a testing technique empirically. With the source code embedded in the system under test, modifying it between test executions for a large number of test cases was simply infeasible. This made a popular approach like mutation analysis very difficult to apply at the code level. How have other researchers evaluated techniques where this is the case?

7.2 Intuition from aggregate information

To begin, we might want to develop some higher level intuition regarding how similar techniques are evaluated in the field. Aggregate statistics and their visualizations presented in the earlier parts of Section 5 can help us quickly identify common characteristics of evaluations performed for similar testing technique types and dimensions.

Looking at the evaluation method distribution for the effectiveness dimension in Fig. 12, we see that over 90% of all effectiveness evaluations were made up of experiments and case studies. Given such a large majority (and in our case the difficulty of performing some analytic evaluation), Fig. 12 gives us a clear indication that our evaluation should most probably be some empirical evaluation in the form of an experiment or case study. Fig. 18 gives us similar information, but considers the testing technique type as well. This figure shows that case studies were somewhat more popular in black-box effectiveness evaluations than they were in white-box evaluations. While experiments were certainly the most common method for evaluating the effectiveness of black-box techniques, many papers also utilized case studies. Thus we would likely choose our evaluation method by reading actual papers evaluating the effectiveness of black-box techniques (see Section 7.3) and by considering whether or not a high level of experimental control is possible.

Another area we might be interested in is how often mutation analysis is utilized in effectiveness evaluations of similar techniques. Figs. 13–16 show us that the proportion of evaluations utilizing mutation analysis remains fairly consistent regardless of evaluation method or testing technique type. For black-box evaluations in particular, Fig. 15 shows that about one-third utilize mutation analysis. Being such a popular technique, we keep it in mind when considering how to evaluate our technique.

7.3 Locating related papers

While aggregate information can give us a quick intuition when it comes to evaluation methods and the use of mutation analysis, it fails to provide a more in-depth understanding of the state of the art in similar testing technique evaluations. We may have many finer-grain questions about how to evaluate our technique or just want to examine papers evaluating similar testing techniques for guidance or inspiration. In our case, we are especially interested in how black-box effectiveness evaluations using mutation analysis are performed when access to the source code is limited. This is where a further understanding of the state of the art is necessary and can be obtained from reading papers performing similar technique evaluations.

Fig. 19 and Table 7 help us to easily locate these papers. As mentioned earlier, Table 7 maps each combination of technique type, evaluation dimension, evaluation method, and mutation analysis affiliation to a set of papers along with the set's cardinality. Due to the large number of combinations and table size, Fig. 19 has been provided as a complementary tool for quickly finding the table row we are interested in. To use the tool, we start at the root node labeled "Technique Type" and work our way down the tree by choosing the category we are interested in at each internal node of the tree. For this case study we are interested in learning about black-box technique evaluations, so we take the right path, labeled "Black-Box," to the *Mutation Analysis Used* internal node. Because we are interested in finding papers that utilize mutation analysis, we then take the left branch to the *Evaluation Dimension* internal node. Finally, our interest in effectiveness evaluations leads us to take the leftmost branch labeled "Effect" and arrive at the leaf node, I. This leaf node represents an identifier for the row in Table 7 containing papers evaluating the effectiveness of black-box testing techniques using mutation analysis.

Given our identifier I, we quickly locate the row labeled I in Table 7 (note identifiers are in alphabetical order and color coded) to find papers we are interested in. Table 7 shows there are 22 experiment and 13 case study papers. We are particularly interested in the evaluations where access to the code may be limited between test executions, so we skim through the set of 35 papers to find such evaluations. This reveals three empirical evaluations, [237], [232], and [233], that we can use to learn how other researchers evaluated the effectiveness of their technique under similar conditions. We see that each of the three evaluations are able to apply some form of mutation analysis without altering the source code between test executions by utilizing model-level mutants for various models. In particular, Aichernig et al. [237] reveal a model-based mutation testing tool for UML models and additionally presents a case study demonstrating how model-based mutation testing can be applied to an industrial measurement device using the tool. By referring to Aichernig et al. [237], we see how we might model our own SUT in UML and utilize model-based mutation analysis to evaluate the effectiveness of our technique.

7.4 Guidelines

After reading through related evaluations, we also may want to consult papers providing higher level guidelines pertaining to our evaluation.

To do so, we refer to Table 3 which lists all of the higher level guideline and proposal papers collected in this mapping study by various categories. Looking through these papers and their summaries, we very quickly gather some valuable insights which will help us plan the evaluation of our testing technique. Ciupa et al. [18] tell us that different techniques may be better suited for finding different types of faults and that the nature of faults found should be considered in testing technique evaluations. Daun et al. [32] suggest that we should consider the experience level of human subjects and should probably apply random selection. Biffl et al. [33] and Briand [346] both do an excellent job of warning us about common threats to the validity of empirical research results. Shull et al. [34] and Carver [35] provide reporting guidelines that will help others replicate our study. Finally, a range of papers in the table present applicable mutation cost reduction techniques we may want to consider.

8. Threats to validity

The main threats to the validity of this study are common to most mapping studies. While systematic, our methods of gathering a set of papers representative of the field under investigation and our methods of mapping them are not immune to these issues.

A major validity concern in systematic mapping studies is that the set of gathered papers fails to include relevant papers in the field. There are a few reasons why this is a threat to the validity of our particular study:

1. *Limited search space*: Relevant papers were only searched for in online databases. Furthermore, our search was only applied to four of the most common online databases. It is possible relevant papers not published online or published in a different online database were missed.

2. *Language barrier*: Only papers written in English were considered in this study. One paper from the initial search was excluded on this basis. It is possible this paper was relevant.

3. *Search string*: The search string chosen obviously has a large impact on the ability of a search to return relevant papers. It is possible the search string used in this study resulted in relevant papers not being returned from online sources. We attempted to mitigate this threat by systematically deriving our search string from our research goal as suggested by Nair et al. [9] and by applying iterative refinements to our search string based on search results (discussed in Section 3.2).

4. *Misleading titles and abstracts*: Some relevant papers may have been excluded in title and abstract exclusion due to titles and abstracts not accurately reflecting the content of papers.

Another major validity concern in systematic mapping studies is that gathered relevant papers are misclassified. This is a concern in our study due to the possibility of author error and poorly written abstracts. The threat is reduced by the fact that full text skimmings were applied to relevant papers to adequately perform some classifications.

9. Conclusions

With the growing demand for high-quality testing techniques it is important that we evaluate them effectively. An understanding of how we currently evaluate techniques and where our evaluations are lacking can give researchers a better idea of how they should evaluate their techniques as well as initiate research to improve technique evaluations. This article provides such an understanding by mapping out the field in a systematic mapping study; illustrating the current state of the art and identifying research gaps. Based on the state of the art we have presented guidelines for how a researcher should evaluate their particular testing technique and have generated a mapping from categories to sets of papers belonging to them; allowing researchers to easily locate papers in the field that they are interested in.

The study also answers nine specific research questions declared in the introduction:

1. *RQ1.1:* The number of papers published annually increased greatly from 2009 to 2011 and has remained about at that level. Since 2011, on average about 35 relevant papers were published per year.

2. *RQ1.2:* Software Testing, Verification and Reliability and the International Symposium on Software Testing and Analysis are the two main publication venues, with 33 and 24 relevant contributions respectively. Other major publication venues include the International Conference on Automated Software Engineering, the International Conference on Software Engineering, the International Conference on Software Testing, the International Symposium on Foundations of Software Engineering, Empirical Software Engineering, and the International Conference on Software Testing, Verification and Validation.

3. *RQ1.3:* A large majority of contributions (87%) are from academia based on author affiliation. Only 13% have authors affiliated with industry.
4. *RQ2.1:* Experiments, case studies, analytic evaluations, and examples are the main methods used for evaluating software testing techniques.
5. *RQ2.2:* Empirical evaluations in the form of experiments make up a very large majority of evaluation methods. Of these, experiments are used quite a bit more. Analytic evaluations and examples are seldom used.
6. *RQ2.3:* Over half of evaluations are of the effectiveness of software testing techniques. 36% evaluate efficiency. A very small remaining proportion of papers evaluate the applicability and scalability of techniques.
7. *RQ2.4* 47% of evaluations were of white-box techniques, 49% of evaluations were of black-box techniques, and 4% of evaluations were of both white-box and black-box techniques.
8. *RQ2.5:* Based on proper experiment and case study methodologies proposed by Vos et al. [14] and Wohlin et al. [11], respectively, evaluations are of relatively low quality.
9. *RQ2.6:* Most of the papers utilized a method to evaluate a software testing technique. Relatively few papers discussed how testing techniques should be evaluated or proposed a method for doing so.
10. *RQ2.7:* Almost 30% of effectiveness evaluations utilized mutation analysis. This percentage is fairly consistent across white-box and black-box testing technique evaluations.

More generally our work concludes that there is a need for research focused on how testing techniques should be evaluated. Most of the empirical evaluations made were of fairly low quality according to proper methodology guidelines. While it is good that many researchers evaluate their techniques, it seems clear the field is lacking more serious testing technique evaluations that are influenced by findings from guideline research. Maturing in this area may greatly enhance our assessment capabilities and as a result further our understanding of the effectiveness, efficiency, scalability, and applicability of software testing techniques.

References

[1] https://www.ibeta.com/historys-most-expensive-software-bugs/, 2018.
[2] K. Petersen, R. Feldt, S. Mujtaba, M. Mattsson, Systematic mapping studies in software engineering, in: Proceedings of the 12th International Conference on Evaluation and Assessment in Software Engineering, BCS Learning & Development Ltd., 2008, pp. 68–77, http://dl.acm.org.du.idm.oclc.org/citation.cfm?id=2227115. 2227123
[3] N. Juristo, A.M. Morena, S. Vegas, Reviewing 25 years of testing technique experiments, Empir. Softw. Eng. 9 (2004) 7–44.

[4] J.E. Gonzalez, N. Juristo, S. Vegas, A systematic mapping study on testing technique experiments: has the situation changed since 2000? in: Proceedings of the 8th ACM/IEEE International Symposium on Empirical Software Engineering and Measurement, ACM, New York, NY, USA, 2014, ISBN: 978-1-4503-2774-9, pp. 3:1–3:4, https://doi.org/10.1145/2652524.2652569.

[5] E. Engstrom, P. Runeson, M. Skoglund, A systematic review on regression test selection techniques. Inf. Softw. Technol. 52 (1) (2010) 14–30, https://doi.org/10.1016/j.infsof.2009.07.001.

[6] Y. Singh, A. Kaur, B. Suri, S. Singhal, Systematic literature review on regression test prioritization techniques, Informatica (Slovenia) 36 (2012) 379–408.

[7] Y. Jia, M. Harman, An analysis and survey of the development of mutation testing. 37 (5) (2011) 649–678, https://doi.org/10.1109/TSE.2010.62.

[8] L.C. Briand, A critical analysis of empirical research in software testing, in: International Symposium on Empirical Software Engineering and Measurement, 2007.

[9] S. Nair, J. Luis de la Vara, M. Sabetzadeh, L. Briand, An extended systematic literature review on provision of evidence for safety certification, Inf. Softw. Technol. 56 (2014) 689–717.

[10] S. Jalali, C. Wohlin, Systematic literature studies: database searches vs. backward snowballing. in: Proceedings of the ACM-IEEE International Symposium on Empirical Software Engineering and Measurement, ACM, 2012, ISBN: 978-1-4503-1056-7, pp. 29–38, https://doi.org/10.1145/2372251.2372257.

[11] C. Wohlin, P. Runeson, M. Hst, M.C. Ohlsson, B. Regnell, A. Wessln, Experimentation in Software Engineering, Springer Publishing Company, Incorporated, 2012.

[12] P. Runeson, M. Host, A. Rainer, B. Regnell, Case Study Research in Software Engineering: Guidelines and Examples, first ed., Wiley Publishing, 2012.

[13] P. Ammann, J. Offutt, Introduction to Software Testing, first, Cambridge University Press, New York, NY, USA, 2008.

[14] T.E.J. Vos, B. Marin, M.J. Escalona, A. Marchetto, A methodological framework for evaluating software testing techniques and tools, in: 2012 12th International Conference on Quality Software, pp. 230–239, 10.1109/QSIC.2012.16.

[15] R. Natella, D. Cotroneo, H.S. Madeira, Assessing dependability with software fault injection: a survey. ACM Comput. Surv. 48 (3) (2016) 44:1–44:55, https://doi.org/10.1145/2841425.

[16] C. Cifuentes, C. Hoermann, N. Keynes, L. Li, S. Long, E. Mealy, M. Mounteney, B. Scholz, BegBunch: benchmarking for C bug detection tools, in: Proceedings of the 2nd International Workshop on Defects in Large Software Systems: Held in Conjunction with the ACM SIGSOFT International Symposium on Software Testing and Analysis (ISSTA 2009), ACM, New York, NY, USA, 2009, ISBN: 978-1-60558-654-0, pp. 16–20.

[17] S.K. Nath, R. Merkel, M.F. Lau, On the improvement of a fault classification scheme with implications for white-box testing. in: Proceedings of the 27th Annual ACM Symposium on Applied Computing, ACM, New York, NY, USA, 2012, ISBN: 978-1-4503-0857-1, pp. 1123–1130, https://doi.org/10.1145/2245276.2231953.

[18] I. Ciupa, A. Pretschner, M. Oriol, A. Leitner, B. Meyer, On the number and nature of faults found by random testing, Softw. Testing Verification Reliab. 21 (1) (2009) 3–28, http://onlinelibrary.wiley.com/doi/abs/10.1002/stvr.415.

[19] T.Y. Chen, R. Merkel, An upper bound on software testing effectiveness. ACM Trans. Softw. Eng. Methodol. 17 (3) (2008) 16:1–16:27, https://doi.org/10.1145/1363102.1363107.

[20] Y. Zhang, A. Mesbah, Assertions are strongly correlated with test suite effectiveness. in: Proceedings of the 2015 10th Joint Meeting on Foundations of Software Engineering, ACM, New York, NY, USA, 2015, ISBN: 978-1-4503-3675-8, pp. 214–224, https://doi.org/10.1145/2786805.2786858.

[21] M. Gligoric, A. Groce, C. Zhang, R. Sharma, M.A. Alipour, D. Marinov, Comparing non-adequate test suites using coverage criteria. in: Proceedings of the 2013 International Symposium on Software Testing and Analysis, ACM, New York, NY, USA, 2013, ISBN: 978-1-4503-2159-4, pp. 302–313, https://doi.org/10.1145/2483760.2483769.

[22] K. Dewey, P. Conrad, M. Craig, E. Morozova, Evaluating test suite effectiveness and assessing student code via constraint logic programming. in: Proceedings of the 2017 ACM Conference on Innovation and Technology in Computer Science Education, ACM, New York, NY, USA, 2017, ISBN: 978-1-4503-4704-4, pp. 317–322, https://doi.org/10.1145/3059009.3059051.

[23] L. Yang, Z. Dang, T.R. Fischer, Information gain of black-box testing, Formal Aspects of Computing 23 (4) (2011) 513–539, http://link.springer.com/article/10.1007/s00165-011-0175-6.

[24] J. Czerwonka, On use of coverage metrics in assessing effectiveness of combinatorial test designs. in: 2013 IEEE Sixth International Conference on Software Testing, Verification and Validation Workshops, March, 2013, pp. 257–266, https://doi.org/10.1109/ICSTW.2013.76.

[25] K. Fawaz, F. Zaraket, W. Masri, H. Harkous, PBCOV: a property-based coverage criterion, Softw. Qual. J. 23 (1) (2015) 171–202, http://link.springer.com/article/10.1007/s11219-014-9237-3.

[26] K. Koster, D. Kao, State coverage: a structural test adequacy criterion for behavior checking. in: The 6th Joint Meeting on European Software Engineering Conference and the ACM SIGSOFT Symposium on the Foundations of Software Engineering: Companion Papers, ACM, New York, NY, USA, 2007, ISBN: 978-1-59593-812-1, pp. 541–544, https://doi.org/10.1145/1295014.1295036.

[27] S.R.S. Souza, S.R. Vergilio, P.S.L. Souza, A.S. Simão, A.C. Hausen, Structural testing criteria for message-passing parallel programs, Concurr Comput: Pract. Exp. 20 (16) (2008) 1893–1916. ISSN: 1532-0634, http://onlinelibrary.wiley.com/doi/abs/10.1002/cpe.1297.

[28] G. Gay, M. Staats, M. Whalen, M.P.E. Heimdahl, The risks of coverage-directed test case generation. IEEE Trans. Softw. Eng. 41 (8) (2015) 803–819, https://doi.org/10.1109/TSE.2015.2421011.

[29] P. Strooper, M.A. Wojcicki, Selecting V V technology combinations: how to pick a winner? in: 12th IEEE International Conference on Engineering Complex Computer Systems (ICECCS 2007) 2007, pp. 87–96, https://doi.org/10.1109/ICECCS.2007.40.

[30] T. Kanstrén, Towards a deeper understanding of test coverage, J. Softw. Maint. Evol. Res. Pract. 20 (1) (2007) 59–76. ISSN: 1532-0618, http://onlinelibrary.wiley.com/doi/abs/10.1002/smr.362.

[31] Y. Guo, S. Sampath, Web application fault classification—an exploratory study. in: Proceedings of the Second ACM-IEEE International Symposium on Empirical Software Engineering and Measurement, ACM, New York, NY, USA, 2008, ISBN: 978-1-59593-971-5, pp. 303–305, https://doi.org/10.1145/1414004.1414060.

[32] M. Daun, A. Salmon, T. Weyer, K. Pohl, The impact of students' skills and experiences on empirical results: a controlled experiment with undergraduate and graduate students. in: Proceedings of the 19th International Conference on Evaluation and Assessment in Software Engineering, ACM, New York, NY, USA, 2015, ISBN: 978-1-4503-3350-4, pp. 29:1–29:6, https://doi.org/10.1145/2745802.2745829.

[33] S. Biffl, M. Kalinowski, F. Ekaputra, A.A. Neto, T. Conte, D. Winkler, Towards a semantic knowledge base on threats to validity and control actions in controlled experiments, in: Proceedings of the 8th ACM/IEEE International Symposium on Empirical Software Engineering and Measurement, ACM, New York, NY, USA, 2014, ISBN: 978-1-4503-2774-9, pp. 49:1–49:4, http://doi.acm.org/10.1145/2652524.2652568.

[34] F.J. Shull, J.C. Carver, S. Vegas, N. Juristo, The role of replications in empirical software engineering. Empir. Softw. Engg. 13 (2) (2008) 211–218, https://doi.org/10. 1007/s10664-008-9060-1.

[35] J.C. Carver, Towards reporting guidelines for experimental replications: a proposal, RESER (2010) 4.

[36] S.U. Farooq, S. Quadri, Empirical evaluation of software testing techniques in an open source fashion. in: Proceedings of the 2nd International Workshop on Conducting Empirical Studies in Industry, ACM, New York, NY, USA, 2014, ISBN: 978-1-4503-2843-2, pp. 21–24, https://doi.org/10.1145/2593690.2593693.

[37] S. Haschemi, S. Weißleder, A generic approach to run mutation analysis, in: Testing–Practice and Research Techniques, Springer, Berlin, Heidelberg, 2010, pp. 155–164, http://link.springer.com/chapter/10.1007/978-3-642-15585-7_15.

[38] A. Gupta, P. Jalote, An approach for experimentally evaluating effectiveness and efficiency of coverage criteria for software testing, Int. J. Softw. Tools Technol. Transfer 10 (2) (2008) 145–160, http://link.springer.com/article/10.1007/s10009-007-0059-5.

[39] R. Just, G.M. Kapfhammer, F. Schweiggert, Do redundant mutants affect the effectiveness and efficiency of mutation analysis? in: Verification and Validation 2012 IEEE Fifth International Conference on Software Testing, 2012, pp. 720–725, https://doi. org/10.1109/ICST.2012.162.

[40] M. Gligoric, V. Jagannath, Q. Luo, D. Marinov, Efficient mutation testing of multi-threaded code, Softw. Testing Verification Reliab. 23 (5) (2012) 375–403, http:// onlinelibrary.wiley.com/doi/abs/10.1002/stvr.1469.

[41] M. Singh, V.M. Srivastava, Extended firm mutation testing: a cost reduction technique for mutation testing. in: 2017 Fourth International Conference on Image Information Processing (ICIIP), December, 2017, pp. 1–6, https://doi.org/10.1109/ICIIP.2017. 8313788.

[42] L. Zhang, D. Marinov, S. Khurshid, Faster mutation testing inspired by test prioritization and reduction. in: Proceedings of the 2013 International Symposium on Software Testing and Analysis, ACM, New York, NY, USA, 2013, ISBN: 978-1-4503-2159-4, pp. 235–245, https://doi.org/10.1145/2483760.2483782.

[43] R. Gopinath, A. Alipour, I. Ahmed, C. Jensen, A. Groce, Measuring effectiveness of mutant sets. in: 2016 IEEE Ninth International Conference on Software Testing, Verification and Validation Workshops (ICSTW), 2016, pp. 132–141, https://doi. org/10.1109/ICSTW.2016.45.

[44] L. Van Phol, N.T. Binh, I. Parissis, Mutants generation for testing Lustre programs. in: Proceedings of the Eighth International Symposium on Information and Communication Technology, ACM, New York, NY, USA, 2017, ISBN: 978-1-4503-5328-1, pp. 425–430, https://doi.org/10.1145/3155133.3155155.

[45] N. Li, M. West, A. Escalona, V.H.S. Durelli, Mutation testing in practice using Ruby. in: 2015 IEEE Eighth International Conference on Software Testing, Verification and Validation Workshops (ICSTW), April, 2015, pp. 1–6, https://doi.org/10. 1109/ICSTW.2015.7107453.

[46] M. Papadakis, Y. Le Traon, Mutation testing strategies using mutant classification. in: Proceedings of the 28th Annual ACM Symposium on Applied Computing, ACM, New York, NY, USA, 2013, ISBN: 978-1-4503-1656-9, pp. 1223–1229, https://doi.org/10.1145/2480362.2480592.

[47] S. Hamimoune, B. Falah, Mutation testing techniques: a comparative study. in: 2016 International Conference on Engineering MIS (ICEMIS), September, 2016, pp. 1–9, https://doi.org/10.1109/ICEMIS.2016.7745368.

[48] R. Just, The major mutation framework: efficient and scalable mutation analysis for Java. in: Proceedings of the 2014 International Symposium on Software Testing

and Analysis, ACM, New York, NY, USA, 2014, ISBN: 978-1-4503-2645-2, pp. 433–436, https://doi.org/10.1145/2610384.2628053.

[49] A.S. Namin, S. Kakarla, The use of mutation in testing experiments and its sensitivity to external threats. in: Proceedings of the 2011 International Symposium on Software Testing and Analysis, ACM, New York, NY, USA, 2011, ISBN: 978-1-4503-0562-4, pp. 342–352, https://doi.org/10.1145/2001420.2001461.

[50] P. Delgado-Pérez, I. Medina-Bulo, M.G. Merayo, Using evolutionary computation to improve mutation testing, in: Advances in Computational Intelligence, Springer, Cham, 2017, ISBN: 978-3-319-59146-9, pp. 381–391, 978-3-319-59147-6. http://link.springer.com/chapter/10.1007/978-3-319-59147-6_33.

[51] M. Papadakis, N. Malevris, An empirical evaluation of the first and second order mutation testing strategies. in: 2010 Third International Conference on Software Testing, Verification, and Validation Workshops, April, 2010, pp. 90–99, https://doi.org/10.1109/ICSTW.2010.50.

[52] M. Polo, M. Piattini, I. García-Rodríguez, Decreasing the cost of mutation testing with second-order mutants, Softw. Testing Verification Reliab. 19 (2) (2008) 111–131. ISSN: 1099-1689, http://onlinelibrary.wiley.com/doi/abs/10.1002/stvr.392.

[53] S. Mirshokraie, A. Mesbah, K. Pattabiraman, Efficient JavaScript mutation testing. in: Verification and Validation 2013 IEEE Sixth International Conference on Software Testing, March, 2013, pp. 74–83, https://doi.org/10.1109/ICST.2013.23.

[54] R.Just, M.D. Ernst, G. Fraser, Efficient mutation analysis by propagating and partitioning infected execution states. in: Proceedings of the 2014 International Symposium on Software Testing and Analysis, ACM, New York, NY, USA, 2014, ISBN: 978-1-4503-2645-2, pp. 315–326, https://doi.org/10.1145/2610384.2610388.

[55] M. Kintis, M. Papadakis, N. Malevris, Evaluating mutation testing alternatives: a collateral experiment. in: 2010 Asia Pacific Software Engineering Conference, 2010, pp. 300–309, https://doi.org/10.1109/APSEC.2010.42.

[56] T. Carwalo, S. Jaswal, Exploring hybrid approach for mutant reduction in software testing. in: 2015 International Conference on Communication, Information Computing Technology (ICCICT), January, 2015, pp. 1–4, https://doi.org/10.1109/ICCICT.2015.7045699.

[57] C. Zhou, P. Frankl, JDAMA: Java database application mutation analyser. Softw. Testing Verification Reliab. 21 (3) (2011) 241–263. ISSN: 1099-1689, https://doi.org/10.1002/stvr.462.

[58] L.T.M. Hanh, N.T. Binh, Mutation operators for Simulink models. in: 2012 Fourth International Conference on Knowledge and Systems Engineering, August, 2012, pp. 54–59, https://doi.org/10.1109/KSE.2012.22.

[59] Y. Khan, J. Hassine, Mutation operators for the Atlas Transformation Language. in: 2013 IEEE Sixth International Conference on Software Testing, Verification and Validation Workshops, March, 2013, pp. 43–52, https://doi.org/10.1109/ICSTW.2013.13.

[60] P.R. Mateo, M.P. Usaola, Parallel mutation testing, Softw. Testing Verification Reliab. 23 (4) (2012) 315–350, http://onlinelibrary.wiley.com/doi/abs/10.1002/stvr.1471.

[61] P.R. Mateo, M.P. Usaola, Reducing mutation costs through uncovered mutants, Softw. Testing Verification Reliab. 25 (5-7) (2014) 464–489. ISSN: 1099-1689, http://onlinelibrary.wiley.com/doi/abs/10.1002/stvr.1534.

[62] M. Gligoric, L. Zhang, C. Pereira, G. Pokam, Selective mutation testing for concurrent code. in: Proceedings of the 2013 International Symposium on Software Testing and Analysis, ACM, New York, NY, USA, 2013, ISBN: 978-1-4503-2159-4, pp. 224–234, https://doi.org/10.1145/2483760.2483773.

[63] Q. Zhu, A. Panichella, A. Zaidman, Speeding-up mutation testing via data compression and state infection. in: 2017 IEEE International Conference on Software Testing,

Verification and Validation Workshops (ICSTW), March, 2017, pp. 103–109, https://doi.org/10.1109/ICSTW.2017.25.

[64] Y.-S. Ma, Y.-R. Kwon, S.-W. Kim, Statistical investigation on class mutation operators, Etri J. 31 (2) (2009) 140–150. ISSN: 2233-7326, http://onlinelibrary.wiley.com/doi/abs/10.4218/etrij.09.0108.0356.

[65] R. Gopinath, C. Jensen, A. Groce, Topsy-Turvy: a smarter and faster parallelization of mutation analysis. in: Proceedings of the 38th International Conference on Software Engineering Companion, ACM, New York, NY, USA, 2016, ISBN: 978-1-4503-4205-6, pp. 740–743, https://doi.org/10.1145/2884781.2892655.

[66] R. Just, G.M. Kapfhammer, F. Schweiggert, Using conditional mutation to increase the efficiency of mutation analysis. in: Proceedings of the 6th International Workshop on Automation of Software Test, ACM, New York, NY, USA, 2011, ISBN: 978-1-4503-0592-1, pp. 50–56, https://doi.org/10.1145/1982595.1982606.

[67] F. Lonetti, E. Marchetti, X-MuT: a tool for the generation of XSLT mutants. in: 2010 Seventh International Conference on the Quality of Information and Communications Technology, September, 2010, pp. 280–285, https://doi.org/10.1109/QUATIC.2010.52.

[68] X. Devroey, G. Perrouin, M. Cordy, M. Papadakis, A. Legay, P.-Y. Schobbens, A variability perspective of mutation analysis. in: Proceedings of the 22nd ACM SIGSOFT International Symposium on Foundations of Software Engineering, ACM, New York, NY, USA, 2014, ISBN: 978-1-4503-3056-5, pp. 841–844, https://doi.org/10.1145/2635868.2666610.

[69] X. Devroey, G. Perrouin, M. Papadakis, A. Legay, P.-Y. Schobbens, P. Heymans, Featured model-based mutation analysis. in: Proceedings of the 38th International Conference on Software Engineering, ACM, New York, NY, USA, 2016, ISBN: 978-1-4503-3900-1, pp. 655–666, https://doi.org/10.1145/2884781.2884821.

[70] L. Cseppento˝, Z. Micskei, Evaluating code-based test input generator tools, Softw. Testing Verification Reliab. 27 (6) (2017) e1627, http://onlinelibrary.wiley.com/doi/abs/10.1002/stvr.1627.

[71] J.M. Rojas, G. Fraser, A. Arcuri, Seeding strategies in search-based unit test generation, Softw. Testing Verification Reliab. 26 (5) (2016) 366–401. ISSN: 1099-1689, http://onlinelibrary.wiley.com/doi/abs/10.1002/stvr.1601.

[72] S.R.S. Souza, P.S.L. Souza, M.A.S. Brito, A.S. Simao, E.J. Zaluska, Empirical evaluation of a new composite approach to the coverage criteria and reachability testing of concurrent programs, Softw. Testing Verification Reliab. 25 (3) (2015) 310–332, http://onlinelibrary.wiley.com/doi/abs/10.1002/stvr.1568.

[73] A. Lakehal, I. Parissis, Structural coverage criteria for LUSTRE/SCADE programs, Softw. Testing Verification Reliab. 19 (2) (2008) 133–154. ISSN: 1099-1689, http://onlinelibrary.wiley.com/doi/abs/10.1002/stvr.394.

[74] E.H. Choi, T. Fujiwara, O. Mizuno, Weighting for combinatorial testing by Bayesian inference. in: 2017 IEEE International Conference on Software Testing, Verification and Validation Workshops (ICSTW), March, 2017, pp. 389–391, https://doi.org/10.1109/ICSTW.2017.73.

[75] M. Patrick, Y. Jia, Kernel density adaptive random testing. in: 2015 IEEE Eighth International Conference on Software Testing, Verification and Validation Workshops (ICSTW), April, 2015, pp. 1–10, https://doi.org/10.1109/ICSTW.2015.7107451.

[76] G. Fraser, M. Staats, P. McMinn, A. Arcuri, F. Padberg, Does automated white-box test generation really help software testers? in: Proceedings of the 2013 International Symposium on Software Testing and Analysis, ACM, New York, NY, USA, 2013, ISBN: 978-1-4503-2159-4, pp. 291–301, https://doi.org/10.1145/2483760.2483774.

[77] A.M.R. Vincenzi, T. Bachiega, D.G. de Oliveira, S.R.S. de Souza, J.C. Maldonado, The complementary aspect of automatically and manually generated test case sets. in: Proceedings of the 7th International Workshop on Automating Test Case Design, Selection, and Evaluation, ACM, New York, NY, USA, 2016, ISBN: 978-1-4503-4401-2, pp. 23–30, https://doi.org/10.1145/2994291.2994295.

[78] P. Ponzio, N. Aguirre, M.F. Frias, W. Visser, Field-exhaustive testing. in: Proceedings of the 2016 24th ACM SIGSOFT International Symposium on Foundations of Software Engineering, ACM, New York, NY, USA, 2016, ISBN: 978-1-4503-4218-6, pp. 908–919, https://doi.org/10.1145/2950290.2950336.

[79] F.C.M. Souza, M. Papadakis, Y. Le Traon, M.E. Delamaro, Strong mutation-based test data generation using hill climbing. in: Proceedings of the 9th International Workshop on Search-Based Software Testing, ACM, New York, NY, USA, 2016, ISBN: 978-1-4503-4166-0, pp. 45–54, https://doi.org/10.1145/2897010.2897012.

[80] M. Papadakis, N. Malevris, M. Kallia, Towards automating the generation of mutation tests. in: Proceedings of the 5th Workshop on Automation of Software Test, ACM, New York, NY, USA, 2010, ISBN: 978-1-60558-970-1, pp. 111–118, https://doi.org/10.1145/1808266.1808283.

[81] N. Li, T. Xie, N. Tillmann, J. de Halleux, W. Schulte, Reggae: automated test generation for programs using complex regular expressions. in: Proceedings of the 2009 IEEE/ACM International Conference on Automated Software Engineering, IEEE Computer Society, Washington, DC, USA, 2009, ISBN: 978-0-7695-3891-4, pp. 515–519, https://doi.org/10.1109/ASE.2009.67.

[82] F. Saglietti, F. Pinte, Automated unit and integration testing for component-based software systems. in: Proceedings of the International Workshop on Security and Dependability for Resource Constrained Embedded Systems, ACM, New York, NY, USA, 2010, ISBN: 978-1-4503-0368-2, pp. 5:1–5:6, https://doi.org/10.1145/1868433.1868440.

[83] W. Zheng, Q. Zhang, M. Lyu, T. Xie, Random unit-test generation with MUT-aware sequence recommendation. in: Proceedings of the IEEE/ACM International Conference on Automated Software Engineering, ACM, New York, NY, USA, 2010, ISBN: 978-1-4503-0116-9, pp. 293–296, https://doi.org/10.1145/1858996.1859054.

[84] H. Wang, W.K. Chan, T.H. Tse, Improving the effectiveness of testing pervasive software via context diversity. ACM Trans. Auton. Adapt. Syst. 9 (2) (2014) 9:1–9:28, https://doi.org/10.1145/2620000.

[85] L. Bajada, M. Micallef, C. Colombo, Using control flow analysis to improve the effectiveness of incremental mutation testing. in: Proceedings of the 14th International Workshop on Principles of Software Evolution, ACM, New York, NY, USA, 2015, ISBN: 978-1-4503-3816-5, pp. 73–78, https://doi.org/10.1145/2804360.2804369.

[86] M. Papadakis, N. Malevris, Automatically performing weak mutation with the aid of symbolic execution, concolic testing and search-based testing, Softw. Qual. J. 19 (4) (2011) 691, http://link.springer.com/article/10.1007/s11219-011-9142-y.

[87] K. Jamrozik, G. Fraser, N. Tillman, J. de Halleux, Generating test suites with augmented dynamic symbolic execution, in: Tests and Proofs, Springer, Berlin, Heidelberg, 2013, ISBN: 978-3-642-38915-3, pp. 152–167, 978-3-642-38916-0. http://link.springer.com/chapter/10.1007/978-3-642-38916-0_9.

[88] E.P. Enoiu, D. Sundmark, A. Čaušević, R. Feldt, P. Pettersson, Mutation-based test generation for PLC embedded software using model checking, in: Testing Software and Systems, Springer, Cham, 2016, ISBN: 978-3-319-47442-7, pp. 155–171, 978-3-319-47443-4. http://link.springer.com/chapter/10.1007/978-3-319-47443-4_10.

[89] K. Kähkönen, R. Kindermann, K. Heljanko, I. Niemelä, Experimental comparison of concolic and random testing for Java card applets, in: Model Checking Software,

Springer, Berlin, Heidelberg, 2010, ISBN: 978-3-642-16163-6, pp. 22–39, 978-3-642-16164-3. http://link.springer.com/chapter/10.1007/978-3-642-16164-3_3.

[90] J. Campos, Y. Ge, G. Fraser, M. Eler, A. Arcuri, An empirical evaluation of evolutionary algorithms for test suite generation, in: Search Based Software Engineering, Springer, Cham, 2017, ISBN: 978-3-319-66298-5, pp. 33–48, 978-3-319-66299-2, http://link.springer.com/chapter/10.1007/978-3-319-66299-2_3.

[91] G. Fraser, A. Zeller, Mutation-driven generation of unit tests and oracles. IEEE Trans. Softw. Eng. 38 (2) (2012) 278–292, https://doi.org/10.1109/TSE.2011.93.

[92] G. Fraser, A. Zeller, Generating parameterized unit tests. in: Proceedings of the 2011 International Symposium on Software Testing and Analysis, ACM, New York, NY, USA, 2011, ISBN: 978-1-4503-0562-4, pp. 364–374, https://doi.org/10.1145/2001420.2001464.

[93] S. Mouchawrab, L.C. Briand, Y. Labiche, M.D. Penta, Assessing, comparing, and combining state machine-based testing and structural testing: a series of experiments. IEEE Trans. Softw. Eng. 37 (2) (2011) 161–187. ISSN: 0098-5589, https://doi.org/10.1109/TSE.2010.32.

[94] J. Tuya, J. Dolado, M.J. Suarez-Cabal, C. de la Riva, A controlled experiment on white-box database testing. SIGSOFT Softw. Eng. Notes 33 (1) (2008) 8:1–8:6, https://doi.org/10.1145/1344452.1344462.

[95] Y. Maezawa, K. Nishiura, H. Washizaki, S. Honiden, Validating Ajax applications using a delay-based mutation technique. in: Proceedings of the 29th ACM/IEEE International Conference on Automated Software Engineering, ACM, New York, NY, USA, 2014, ISBN: 978-1-4503-3013-8, pp. 491–502, https://doi.org/10.1145/2642937.2642996.

[96] H. Yoshida, S. Tokumoto, M.R. Prasad, I. Ghosh, T. Uehara, FSX: fine-grained incremental unit test generation for C/C++ programs. in: Proceedings of the 25th International Symposium on Software Testing and Analysis, ACM, New York, NY, USA, 2016, ISBN: 978-1-4503-4390-9, pp. 106–117, https://doi.org/10.1145/2931037.2931055.

[97] F. Pinte, N. Oster, F. Saglietti, Techniques and tools for the automatic generation of optimal test data at code, model and interface level. in: Companion of the 30th International Conference on Software Engineering, ACM, New York, NY, USA, 2008, ISBN: 978-1-60558-079-1, pp. 927–928, https://doi.org/10.1145/1370175.1370191.

[98] A. Dreyfus, P.-C. Héam, O. Kouchnarenko, C. Masson, A random testing approach using pushdown automata, Softw. Testing Verification Reliab. 24 (8) (2014) 656–683, http://onlinelibrary.wiley.com/doi/abs/10.1002/stvr.1526.

[99] M. Kim, Y. Kim, Y. Choi, Concolic testing of the multi-sector read operation for flash storage platform software, Formal Aspects of Computing 24 (3) (2012) 355–374, http://link.springer.com/article/10.1007/s00165-011-0200-9.

[100] S.U. Farooq, S.M.K. Quadri, N. Ahmad, A replicated empirical study to evaluate software testing methods, J. Softw. Evol. Process 29 (9) (2017) e1883, http://onlinelibrary.wiley.com/doi/abs/10.1002/smr.1883.

[101] A. Arcuri, It really does matter how you normalize the branch distance in search-based software testing, Softw. Testing Verification Reliab. 23 (2) (2011) 119–147, http://onlinelibrary.wiley.com/doi/abs/10.1002/stvr.457.

[102] J. Malburg, G. Fraser, Search-based testing using constraint-based mutation. Softw. Testing, Verification and Reliability 24 (6) (2013) 472–495. ISSN: 1099-1689, https://doi.org/10.1002/stvr.1508.

[103] H. Wang, J. Xing, Q. Yang, W. Song, X. Zhang, Generating effective test cases based on satisfiability modulo theory solvers for service-oriented workflow applications, Softw. Testing Verification Reliab. 26 (2) (2015) 149–169, http://onlinelibrary.wiley.com/doi/abs/10.1002/stvr.1592.

[104] R.T. Alexander, J. Offutt, A. Stefik, Testing coupling relationships in object-oriented programs, Softw. Testing Verification Reliab. 20 (4) (2010) 291–327, http://onlinelibrary.wiley.com/doi/abs/10.1002/stvr.417.

[105] S. Godboley, D.P. Mohapatra, A. Das, R. Mall, An improved distributed concolic testing approach, Softw. Pract. Exp. 47 (2) (2016) 311–342, http://onlinelibrary.wiley.com/doi/abs/10.1002/spe.2405.

[106] J.H. Hayes, I.R. Chemannoor, E.A. Holbrook, Improved code defect detection with fault links, Softw. Testing Verification Reliab. 21 (4) (2010) 299–325, http://onlinelibrary.wiley.com/doi/abs/10.1002/stvr.426.

[107] T. Do, A.C.M. Fong, R. Pears, Scalable automated test generation using coverage guidance and random search. in: 2012 7th International Workshop on Automation of Software Test (AST), June, 2012, pp. 71–75, https://doi.org/10.1109/IWAST.2012.6228993.

[108] M. Parthiban, M.R. Sumalatha, GASE—an input domain reduction and branch coverage system based on genetic algorithm and symbolic execution. in: 2013 International Conference on Information Communication and Embedded Systems (ICICES), February, 2013, pp. 429–433, https://doi.org/10.1109/ICICES.2013.6508273.

[109] S. Godboley, A. Dutta, B. Besra, D.P. Mohapatra, Green-JEXJ: a new tool to measure energy consumption of improved concolic testing. in: 2015 International Conference on Green Computing and Internet of Things (ICGCIoT), 2015, pp. 36–41, https://doi.org/10.1109/ICGCIoT.2015.7380424.

[110] A. Arcuri, RESTful API automated test case generation. in: 2017 IEEE International Conference on Software Quality, Reliability and Security (QRS), July, 2017, pp. 9–20, https://doi.org/10.1109/QRS.2017.11.

[111] T. Guo, P. Zhang, X. Wang, Q. Wei, GramFuzz: Fuzzing testing of web browsers based on grammar analysis and structural mutation. in: 2013 Second International Conference on Informatics Applications (ICIA), September, 2013, pp. 212–215, https://doi.org/10.1109/ICoIA.2013.6650258.

[112] L. Bentes, H. Rocha, E. Valentin, R. Barreto, JFORTES: Java formal unit TESt generation. in: 2016 VI Brazilian Symposium on Computing Systems Engineering (SBESC), November, 2016, pp. 16–23, https://doi.org/10.1109/SBESC.2016.012.

[113] A. Andalib, S.M. Babamir, A new approach for test case generation by discrete particle swarm optimization algorithm. in: 2014 22nd Iranian Conference on Electrical Engineering (ICEE), May, 2014, pp. 1180–1185, https://doi.org/10.1109/IranianCEE.2014.6999714.

[114] G. Fraser, A. Arcuri, The seed is strong: seeding strategies in search-based software testing. in: Verification and Validation 2012 IEEE Fifth International Conference on Software Testing, 2012, pp. 121–130, https://doi.org/10.1109/ICST.2012.92.

[115] Y. Wang, Y. Wang, Use neural network to improve fault injection testing. in: 2017 IEEE International Conference on Software Quality, Reliability and Security Companion (QRS-C), July, 2017, pp. 377–384, https://doi.org/10.1109/QRS-C.2017.69.

[116] N. Juristo, S. Vegas, M. Solari, S. Abrahao, I. Ramos, Comparing the effectiveness of equivalence partitioning, branch testing and code reading by stepwise abstraction applied by subjects. in: Verification and Validation 2012 IEEE Fifth International Conference on Software Testing, 2012, pp. 330–339, https://doi.org/10.1109/ICST.2012.113.

[117] S. Poulding, J.A. Clark, H. Waeselynck, A principled evaluation of the effect of directed mutation on search-based statistical testing. in: 2011 IEEE Fourth International Conference on Software Testing, Verification and Validation Workshops, March, 2011, pp. 184–193, https://doi.org/10.1109/ICSTW.2011.36.

[118] S. Godboley, A. Dutta, D.P. Mohapatra, Java-HCT: an approach to increase MC/DC using hybrid concolic testing for Java programs, in: 2016 Federated Conference on Computer Science and Information Systems (FedCSIS), 2016, pp. 1709–1713 september.

[119] R. Jin, S. Jiang, H. Zhang, Generation of test data based on genetic algorithms and program dependence analysis. in: 2011 IEEE International Conference on Cyber Technology in Automation, Control, and Intelligent Systems, 2011, pp. 116–121, https://doi.org/10.1109/CYBER.2011.6011775.

[120] M. Deng, R. Chen, Z. Du, Automatic test data generation model by combining dataflow analysis with genetic algorithm. in: 2009 Joint Conferences on Pervasive Computing (JCPC), December, 2009, pp. 429–434, https://doi.org/10.1109/JCPC.2009.5420148.

[121] G. Gay, The fitness function for the job: search-based generation of test suites that detect real faults. in: 2017 IEEE International Conference on Software Testing, Verification and Validation (ICST), 2017, pp. 345–355, https://doi.org/10.1109/ICST.2017.38 march.

[122] W. Zhang, J. Lim, R. Olichandran, J. Scherpelz, G. Jin, S. Lu, T. Reps, ConSeq: detecting concurrency bugs through sequential errors. in: Proceedings of the Sixteenth International Conference on Architectural Support for Programming Languages and Operating Systems, ACM, New York, NY, USA, 2011, ISBN: 978-1-4503-0266-1, pp. 251–264, https://doi.org/10.1145/1950365.1950395.

[123] K. Mao, M. Harman, Y. Jia, Sapienz: multi-objective automated testing for Android applications. in: Proceedings of the 25th International Symposium on Software Testing and Analysis, ACM, New York, NY, USA, 2016, ISBN: 978-1-4503-4390-9, pp. 94–105, https://doi.org/10.1145/2931037.2931054.

[124] L.S. Silva, M. van Someren, Evolutionary testing of object-oriented software. in: Proceedings of the 2010 ACM Symposium on Applied Computing, ACM, New York, NY, USA, 2010, ISBN: 978-1-60558-639-7, pp. 1126–1130, https://doi.org/10.1145/1774088.1774326.

[125] K. Agarwal, G. Srivastava, Towards software test data generation using discrete quantum particle swarm optimization. in: Proceedings of the 3rd India Software Engineering Conference, ACM, New York, NY, USA, 2010, ISBN: 978-1-60558-922-0, pp. 65–68, https://doi.org/10.1145/1730874.1730888.

[126] P. Godefroid, A. Kiezun, M.Y. Levin, Grammar-based whitebox fuzzing. in: Proceedings of the 29th ACM SIGPLAN Conference on Programming Language Design and Implementation, ACM, New York, NY, USA, 2008, ISBN: 978-1-59593-860-2, pp. 206–215, https://doi.org/10.1145/1375581.1375607.

[127] M. Mirzaaghaei, A. Mesbah, DOM-based test adequacy criteria for web applications. in: Proceedings of the 2014 International Symposium on Software Testing and Analysis, ACM, New York, NY, USA, 2014, ISBN: 978-1-4503-2645-2, pp. 71–81, https://doi.org/10.1145/2610384.2610406.

[128] Y. Zou, Z. Chen, Y. Zheng, X. Zhang, Z. Gao, Virtual DOM coverage for effective testing of dynamic Web applications. in: Proceedings of the 2014 International Symposium on Software Testing and Analysis, ACM, New York, NY, USA, 2014, ISBN: 978-1-4503-2645-2, pp. 60–70, https://doi.org/10.1145/2610384.2610399.

[129] D. Amalfitano, N. Amatucci, A.R. Fasolino, P. Tramontana, AGRippin: a novel search based testing technique for Android applications. in: Proceedings of the 3rd International Workshop on Software Development Lifecycle for Mobile, ACM, New York, NY, USA, 2015, ISBN: 978-1-4503-3815-8, pp. 5–12, https://doi.org/10.1145/2804345.2804348.

[130] W.G.J. Halfond, A. Orso, Improving test case generation for web applications using automated interface discovery. in: Proceedings of the 6th Joint Meeting of the

European Software Engineering Conference and the ACM SIGSOFT Symposium on The Foundations of Software Engineering, ACM, New York, NY, USA, 2007, ISBN: 978-1-59593-811-4, pp. 145–154, https://doi.org/10.1145/1287624.1287646.

[131] A. Windisch, S. Wappler, J. Wegener, Applying particle swarm optimization to software testing. in: Proceedings of the 9th Annual Conference on Genetic and Evolutionary Computation, ACM, New York, NY, USA, 2007, ISBN: 978-1-59593-697-4, pp. 1121–1128, https://doi.org/10.1145/1276958.1277178.

[132] M.A. Alipour, A. Groce, R. Gopinath, A. Christi, Generating focused random tests using directed swarm testing. in: Proceedings of the 25th International Symposium on Software Testing and Analysis, ACM, New York, NY, USA, 2016, ISBN: 978-1-4503-4390-9, pp. 70–81, https://doi.org/10.1145/2931037.2931056.

[133] B.M. Padmanabhuni, H. Beng Kuan Tan, Light-weight rule-based test case generation for detecting buffer overflow vulnerabilities, in: Proceedings of the 10th International Workshop on Automation of Software Test, IEEE Press, Piscataway, NJ, USA, 2015, pp. 48–52, http://dl.acm.org/citation.cfm?id=2819261.2819276.

[134] K. Taneja, Y. Zhang, T. Xie, MODA: automated test generation for database applications via Mock objects. in: Proceedings of the IEEE/ACM International Conference on Automated Software Engineering, ACM, New York, NY, USA, 2010, ISBN: 978-1-4503-0116-9, pp. 289–292, https://doi.org/10.1145/1858996.1859053.

[135] C. Sung, M. Kusano, N. Sinha, C. Wang, Static DOM event dependency analysis for testing web applications. in: Proceedings of the 2016 24th ACM SIGSOFT International Symposium on Foundations of Software Engineering, ACM, New York, NY, USA, 2016, ISBN: 978-1-4503-4218-6, pp. 447–459, https://doi.org/10.1145/2950290.2950292.

[136] N. Chen, S. Kim, Puzzle-based automatic testing: bringing humans into the loop by solving puzzles. in: Proceedings of the 27th IEEE/ACM International Conference on Automated Software Engineering, ACM, New York, NY, USA, 2012, ISBN: 978-1-4503-1204-2, pp. 140–149, https://doi.org/10.1145/2351676.2351697.

[137] T. Avgerinos, A. Rebert, S.K. Cha, D. Brumley, Enhancing symbolic execution with veritesting. in: Proceedings of the 36th International Conference on Software Engineering, ACM, New York, NY, USA, 2014, ISBN: 978-1-4503-2756-5, pp. 1083–1094, https://doi.org/10.1145/2568225.2568293.

[138] S. Monpratarnchai, S. Fujiwara, A. Katayama, T. Uehara, Automated testing for Java programs using JPF-based test case generation. SIGSOFT Softw. Eng. Notes 39 (1) (2014) 1–5. ISSN: 0163-5948, https://doi.org/10.1145/2557833.2560575.

[139] G. Fraser, A. Arcuri, P. McMinn, Test suite generation with memetic algorithms. in: Proceedings of the 15th Annual Conference on Genetic and Evolutionary Computation, ACM, New York, NY, USA, 2013, ISBN: 978-1-4503-1963-8, pp. 1437–1444, https://doi.org/10.1145/2463372.2463548.

[140] J.H. Andrews, F.C.H. Li, T. Menzies, Nighthawk: a two-level genetic-random unit test data generator. in: Proceedings of the Twenty-Second IEEE/ACM International Conference on Automated Software Engineering, ACM, New York, NY, USA, 2007, ISBN: 978-1-59593-882-4, pp. 144–153, https://doi.org/10.1145/1321631.1321654.

[141] A. Farzan, A. Holzer, N. Razavi, H. Veith, Con2Colic Testing. in: Proceedings of the 2013 9th Joint Meeting on Foundations of Software Engineering, ACM, New York, NY, USA, 2013, ISBN: 978-1-4503-2237-9, pp. 37–47, https://doi.org/10.1145/2491411.2491453.

[142] G. Li, P. Li, G. Sawaya, G. Gopalakrishnan, I. Ghosh, S.P. Rajan, GKLEE: concolic verification and test generation for GPUs. in: Proceedings of the 17th ACM SIGPLAN Symposium on Principles and Practice of Parallel Programming, ACM, New York, NY, USA, 2012, ISBN: 978-1-4503-1160-1, pp. 215–224, https://doi.org/10.1145/2145816.2145844.

[143] S. Guo, M. Kusano, C. Wang, Z. Yang, A. Gupta, Assertion guided symbolic execution of multithreaded programs. in: Proceedings of the 2015 10th Joint Meeting on Foundations of Software Engineering, ACM, New York, NY, USA, 2015, ISBN: 978-1-4503-3675-8, pp. 854–865, https://doi.org/10.1145/2786805.2786841.

[144] M. Islam, C. Csallner, Generating test cases for programs that are coded against interfaces and annotations. ACM Trans. Softw. Eng. Methodol. 23 (3) (2014) 21:1–21:38, https://doi.org/10.1145/2544135.

[145] A. Marchetto, F. Ricca, P. Tonella, A case study-based comparison of web testing techniques applied to AJAX web applications, Int. J. Softw. Tools Technol. Transfer 10 (6) (2008) 477–492, http://link.springer.com/article/10.1007/s10009-008-0086-x.

[146] D. Hao, L. Zhang, M.-H. Liu, H. Li, J.-S. Sun, Test-data generation guided by static defect detection, J. Comput. Sci. Technol. 24 (2) (2009) 284–293, http://link.springer.com/article/10.1007/s11390-009-9224-5.

[147] J. Itkonen, M.V. Mäntylä, Are test cases needed? Replicated comparison between exploratory and test-case-based software testing, Empir. Softw. Eng. 19 (2) (2014) 303–342, http://link.springer.com/article/10.1007/s10664-013-9266-8.

[148] W. Afzal, A.N. Ghazi, J. Itkonen, R. Torkar, A. Andrews, K. Bhatti, An experiment on the effectiveness and efficiency of exploratory testing, Empir. Softw. Eng. 20 (3) (2015) 844–878, http://link.springer.com/article/10.1007/s10664-014-9301-4.

[149] P. Braione, G. Denaro, A. Mattavelli, M. Vivanti, A. Muhammad, Software testing with code-based test generators: data and lessons learned from a case study with an industrial software component, Softw. Qual. J. 22 (2) (2014) 311–333, http://link.springer.com/article/10.1007/s11219-013-9207-1.

[150] C. Apa, O. Dieste, E.G. Espinosa G, E.R. Fonseca C, Effectiveness for detecting faults within and outside the scope of testing techniques: an independent replication, Empir. Softw. Eng. 19 (2) (2014) 378–417, http://link.springer.com/article/10.1007/s10664-013-9267-7.

[151] P. Chawla, I. Chana, A. Rana, A novel strategy for automatic test data generation using soft computing technique, Frontiers of Computer Science 9 (3) (2015) 346–363, http://link.springer.com/article/10.1007/s11704-014-3496-9.

[152] C. Mao, Harmony search-based test data generation for branch coverage in software structural testing, Neural Comput. Appl. 25 (1) (2014) 199–216, http://link.springer.com/article/10.1007/s00521-013-1474-z.

[153] D. Gong, Y. Zhang, Generating test data for both path coverage and fault detection using genetic algorithms, Front. Comput. Sci. 7 (6) (2013) 822–837, http://link.springer.com/article/10.1007/s11704-013-3024-3.

[154] S. Godboley, A. Dutta, D.P. Mohapatra, A. Das, R. Mall, Making a concolic tester achieve increased MC/DC, Innovations in Systems and Software Engineering 12 (4) (2016) 319–332, http://link.springer.com/article/10.1007/s11334-016-0284-8.

[155] G. Fraser, A. Arcuri, Achieving scalable mutation-based generation of whole test suites, Empir. Softw. Eng. 20 (3) (2015) 783–812, http://link.springer.com/article/10.1007/s10664-013-9299-z.

[156] C. Engel, R. Hähnle, Generating unit tests from formal proofs, in: Tests and Proofs, Springer, Berlin, Heidelberg, 2007, ISBN: 978-3-540-73769-8, pp. 169–188, 978-3-540-73770-4. http://link.springer.com/chapter/10.1007/978-3-540-73770-4_10.

[157] J. Huo, B. Xue, L. Shang, M. Zhang, Genetic programming for multi-objective test data generation in search based software testing, in: AI 2017: Advances in Artificial Intelligence, Springer, Cham, 2017, ISBN: 978-3-319-63003-8, pp. 169–181, 978-3-319-63004-5 http://link.springer.com/chapter/10.1007/978-3-319-63004-5_14.

[158] Y.-H. Jia, W.-N. Chen, J. Zhang, J.-J. Li, Generating software test data by particle swarm optimization, in: Simulated Evolution and Learning, Springer, Cham, 2014, ISBN: 978-3-319-13562-5, pp. 37–47, 978-3-319-13563-2. http://link.springer.com/chapter/10.1007/978-3-319-13563-2_4.

[159] Y. Hu, H. Jiang, Effective test case generation via concolic execution, in: Proceedings of the 2012 International Conference on Information Technology and Software Engineering, Springer, Berlin, Heidelberg, 2013, ISBN: 978-3-642-34530-2, pp. 157–164, 978-3-642-34531-9. http://link.springer.com/chapter/10.1007/978-3-642-34531-9_17.

[160] A. Gargantini, P. Vavassori, Hardware and Software: Verification and Testing, Springer, Cham, 2014, ISBN: 978-3-319-13337-9, pp. 220–235, 978-3-319-13338-6http://link.springer.com/chapter/10.1007/978-3-319-13338-6_17.

[161] A.M. Bidgoli, H. Haghighi, T.Z. Nasab, H. Sabouri, Using swarm intelligence to generate test data for covering prime paths, in: Fundamentals of Software Engineering, Springer, Cham, 2017, ISBN: 978-3-319-68971-5, pp. 132–147, 978-3-319-68972-2, http://link.springer.com/chapter/10.1007/978-3-319-68972-2_9.

[162] N. Bhattacharya, A. Sakti, G. Antoniol, Y.-G. Gu'eheneuc, G. Pesant, Divide-by-zero exception raising via branch coverage, in: Search Based Software Engineering, Springer, Berlin, Heidelberg, 2011, ISBN: 978-3-642-23715-7, pp. 204–218, 978-3-642-23716-4, http://link.springer.com/chapter/10.1007/978-3-642-23716-4_19.

[163] Y. Zhou, T. Sugihara, Y. Sato, Applying GA with Tabu list for automatically generating test cases based on formal specification, in: Structured Object-Oriented Formal Language and Method, November, Springer, Cham, 2014, ISBN: 978-3-319-17403-7, pp. 17–31, 978-3-319-17404-4, http://link.springer.com/chapter/10.1007/978-3-319-17404-4_2.

[164] G. Fraser, A. Arcuri, A large-scale evaluation of automated unit test generation using evosuite. ACM Trans. Softw. Eng. Methodol. 24 (2) (2014) 8:1–8:42, https://doi.org/10.1145/2685612.

[165] S.Y. Lee, H.J. Choi, Y.J. Jeong, T.H. Kim, H.S. Chae, C.K. Chang, An improved technique of fitness evaluation for evolutionary testing. in: 2011 IEEE 35th Annual Computer Software and Applications Conference Workshops, 2011, pp. 190–193, https://doi.org/10.1109/COMPSACW.2011.41.

[166] N. Alshahwan, M. Harman, Automated web application testing using search based software engineering. in: 2011 26th IEEE/ACM International Conference on Automated Software Engineering (ASE 2011), November, 2011, pp. 3–12, https://doi.org/10.1109/ASE.2011.6100082.

[167] M. Kim, Y. Kim, H. Kim, A comparative study of software model checkers as unit testing tools: an industrial case study. IEEE Trans. Softw. Eng. 37 (2) (2011) 146–160, https://doi.org/10.1109/TSE.2010.68.

[168] K. Liaskos, M. Roper, Automatic test-data generation: an immunological approach. in: Testing: Academic and Industrial Conference Practice and Research Techniques–MUTATION (TAICPART-MUTATION 2007), 2007, pp. 77–81, https://doi.org/10.1109/TAIC.PART.2007.24.

[169] T. Allwood, C. Cadar, S. Eisenbach, High coverage testing of haskell programs. in: Proceedings of the 2011 International Symposium on Software Testing and Analysis, ACM, New York, NY, USA, 2011, ISBN: 978-1-4503-0562-4, pp. 375–385, https://doi.org/10.1145/2001420.2001465.

[170] S. Artzi, A. Kiezun, J. Dolby, F. Tip, D. Dig, A. Paradkar, M.D. Ernst, Finding bugs in dynamic Web applications. in: Proceedings of the 2008 International Symposium on Software Testing and Analysis, ACM, New York, NY, USA, 2008, ISBN: 978-1-60558-050-0, pp. 261–272, https://doi.org/10.1145/1390630.1390662.

[171] S. Zhang, H. Lü, M.D. Ernst, Finding errors in multithreaded GUI applications. in: Proceedings of the 2012 International Symposium on Software Testing and

Analysis, ACM, New York, NY, USA, 2012, ISBN: 978-1-4503-1454-1, pp. 243–253, https://doi.org/10.1145/2338965.2336782.

[172] P.D. Marinescu, G. Candea, Efficient testing of recovery code using fault injection. ACM Trans. Comput. Syst. 29 (4) (2011) 11:1–11:38. ISSN: 0734-2071, https://doi.org/10.1145/2063509.2063511.

[173] I. Ghosh, N. Shafiei, G. Li, W.-F. Chiang, JST: an automatic test generation tool for industrial java applications with strings, in: Proceedings of the 2013 International Conference on Software Engineering, IEEE Press, Piscataway, NJ, USA, 2013, ISBN: 978-1-4673-3076-3, pp. 992–1001, http://dl.acm.org/citation.cfm?id=2486788.2486925.

[174] O. Shacham, N. Bronson, A. Aiken, M. Sagiv, M. Vechev, E. Yahav, Testing atomicity of composed concurrent operations. in: Proceedings of the 2011 ACM International Conference on Object Oriented Programming Systems Languages and Applications, ACM, New York, NY, USA, 2011, ISBN: 978-1-4503-0940-0, pp. 51–64, https://doi.org/10.1145/2048066.2048073.

[175] G. Fraser, A. Arcuri, Sound empirical evidence in software testing, in: Proceedings of the 34th International Conference on Software Engineering, IEEE Press, Piscataway, NJ, USA, 2012, ISBN: 978-1-4673-1067-3, pp. 178–188, http://dl.acm.org/citation.cfm?id=2337223.2337245.

[176] E. Bounimova, P. Godefroid, D. Molnar, Billions and billions of constraints: whitebox Fuzz testing in production, in: Proceedings of the 2013 International Conference on Software Engineering, IEEE Press, Piscataway, NJ, USA, 2013, ISBN: 978-1-4673-3076-3, pp. 122–131, http://dl.acm.org/citation.cfm?id=2486788.2486805.

[177] Z. Rakamarić, STORM: static unit checking of concurrent programs. in: Proceedings of the 32nd ACM/IEEE International Conference on Software Engineering - Volume 2, ACM, New York, NY, USA, 2010, ISBN: 978-1-60558-719-6, pp. 519–520, https://doi.org/10.1145/1810295.1810460.

[178] B. Pasternak, S. Tyszberowicz, A. Yehudai, GenUTest: a unit test and mock aspect generation tool, Int. J. Softw. Tools Technol. Transfer 11 (4) (2009) 273, http://link.springer.com/article/10.1007/s10009-009-0115-4.

[179] N Tillmann,J. de Halleux, Pex-white box test generation for .NET, in: Tests and Proofs, Springer, Berlin, Heidelberg, 2008, ISBN: 978-3-540-79123-2, pp. 134–153, 978-3-540-79124-9, http://linkspringer.com/chapter/10.1007/978-3-540-79124-9_10.

[180] S. Benli, A. Habash, A. Herrmann, T. Loftis, D. Simmonds, A comparative evaluation of unit testing techniques on a mobile platform. in: 2012 Ninth International Conference on Information Technology - New Generations, April, 2012, pp. 263–268, https://doi.org/10.1109/ITNG.2012.45.

[181] D.N. Thi, V.D. Hieu, N.V. Ha, A technique for generating test data using genetic algorithm. in: 2016 International Conference on Advanced Computing and Applications (ACOMP), November, 2016, pp. 67–73, https://doi.org/10.1109/ACOMP.2016.019.

[182] A. Giantsios, N. Papaspyrou, K. Sagonas, Concolic testing for functional languages. in: Proceedings of the 17th International Symposium on Principles and Practice of Declarative Programming, ACM, New York, NY, USA, 2015, ISBN: 978-1-4503-3516-4, pp. 137–148, https://doi.org/10.1145/2790449.2790519.

[183] C. Song, A. Porter, J.S. Foster, iTree: efficiently discovering high-coverage configurations using interaction trees. IEEE Trans. Softw. Eng. 40 (3) (2014) 251–265. ISSN: 0098-5589, https://doi.org/10.1109/TSE.2013.55.

[184] M. Böhme, S. Paul, On the efficiency of automated testing. in: Proceedings of the 22nd ACM SIGSOFT International Symposium on Foundations of Software Engineering, ACM, New York, NY, USA, 2014, ISBN: 978-1-4503-3056-5, pp. 632–642, https://doi.org/10.1145/2635868.2635923.

[185] B.S. Apilli, Fault-based combinatorial testing of Web services. in: Proceedings of the 24th ACM SIGPLAN Conference Companion on Object Oriented Programming

Systems Languages and Applications, ACM, New York, NY, USA, 2009, ISBN: 978-1-60558-768-4, pp. 731–732, https://doi.org/10.1145/1639950.1639987.

[186] P. Godefroid, Higher-order test generation. in: Proceedings of the 32nd ACM SIGPLAN Conference on Programming Language Design and Implementation, ACM, New York, NY, USA, 2011, ISBN: 978-1-4503-0663-8, pp. 258–269, https://doi.org/10.1145/1993498.1993529.

[187] J. Ferrer, F. Chicano, E. Alba, Evolutionary algorithms for the multi-objective test data generation problem, Software: Practice and Experience 42 (11) (2011) 1331–1362, http://onlinelibrary.wiley.com/doi/abs/10.1002/spe.1135.

[188] Z. Zhu, X. Xu, L. Jiao, Improved evolutionary generation of test data for multiple paths in search-based software testing. in: 2017 IEEE Congress on Evolutionary Computation (CEC), June, 2017, pp. 612–620, https://doi.org/10.1109/CEC.2017.7969367.

[189] R. Feldt, S. Poulding, Broadening the Search in search-based software testing: it need not Be evolutionary. in: 2015 IEEE/ACM 8th International Workshop on Search-Based Software Testing, May, 2015, pp. 1–7, https://doi.org/10.1109/SBST.2015.8.

[190] D. Kroening, L. Liang, T. Melham, P. Schrammel, M. Tautschnig, Effective verification of low-level software with nested interrupts. in: 2015 Design, Automation Test in Europe Conference Exhibition (DATE), 2015, pp. 229–234, https://doi.org/10.7873/DATE.2015.0360. march.

[191] A. Baars, M. Harman, Y. Hassoun, K. Lakhotia, P. McMinn, P. Tonella, T. Vos, Symbolic search-based testing. in: 2011 26th IEEE/ACM International Conference on Automated Software Engineering (ASE 2011), November, 2011, pp. 53–62, https://doi.org/10.1109/ASE.2011.6100119.

[192] D. Shannon, I. Ghosh, S. Rajan, S. Khurshid, Efficient symbolic execution of strings for validating Web applications. in: Proceedings of the 2nd International Workshop on Defects in Large Software Systems: Held in Conjunction with the ACM SIGSOFT International Symposium on Software Testing and Analysis (ISSTA 2009), ACM, New York, NY, USA, 2009, ISBN: 978-1-60558-654-0, pp. 22–26, https://doi.org/10.1145/1555860.1555868.

[193] A. Rathore, A. Bohara, R.G. Prashil, T.S.L. Prashanth, P.R. Srivastava, Application of genetic algorithm and tabu search in software testing. in: Proceedings of the Fourth Annual ACM Bangalore Conference, ACM, New York, NY, USA, 2011, ISBN: 978-1-4503-0750-5, pp. 23:1–23:4, https://doi.org/10.1145/1980422.1980445.

[194] K.E.Coons, S. Burckhardt, M. Musuvathi, GAMBIT: effective unit testing for concurrency libraries. in: Proceedings of the 15th ACM SIGPLAN Symposium on Principles and Practice of Parallel Programming, ACM, New York, NY, USA, 2010, ISBN: 978-1-60558-877-3, pp. 15–24, https://doi.org/10.1145/1693453.1693458.

[195] M. Staats, C. Pa˘sa˘reanu, Parallel symbolic execution for structural test generation. in: Proceedings of the 19th International Symposium on Software Testing and Analysis, ACM, New York, NY, USA, 2010, ISBN: 978-1-60558-823-0, pp. 183–194, https://doi.org/10.1145/1831708.1831732.

[196] M. Lin, X. Hou, R. Liu, L. Ge, Enhancing constraint based test generation by local search. in: Proceedings of the 6th International Conference on Software and Computer Applications, ACM, New York, NY, USA, 2017, ISBN: 978-1-4503-4857-7, pp. 154–158, https://doi.org/10.1145/3056662.3056672.

[197] S. Park, B.M.M. Hossain, I. Hussain, C. Csallner, M. Grechanik, K. Taneja, C. Fu, Q. Xie, CarFast: achieving higher statement coverage faster. in: Proceedings of the ACM SIGSOFT 20th International Symposium on the Foundations of Software Engineering, ACM, New York, NY, USA, 2012, ISBN: 978-1-4503-1614-9, pp. 35:1–35:11, https://doi.org/10.1145/2393596.2393636.

[198] K. Kähkönen, O. Saarikivi, K. Heljanko, Using unfoldings in automated testing of multithreaded programs. in: Proceedings of the 27th IEEE/ACM International Conference on Automated Software Engineering, ACM, New York, NY, USA, 2012, ISBN: 978-1-4503-1204-2, pp. 150–159, https://doi.org/10.1145/2351676. 2351698.

[199] J.H. Siddiqui, S. Khurshid, Scaling symbolic execution using staged analysis, Innov. Syst. Softw. Eng. 9 (2) (2013) 119–131, http://link.springer.com/article/10.1007/s11334-013-0196-9.

[200] M. Alshraideh, B.A. Mahafzah, S. Al-Sharaeh, A multiple-population genetic algorithm for branch coverage test data generation, Softw. Qual. J. 19 (3) (2011) 489–513, http://link.springer.com/article/10.1007/s11219-010-9117-4.

[201] T. Tian, D. Gong, Test data generation for path coverage of message-passing parallel programs based on co-evolutionary genetic algorithms, Automated Software Engineering 23 (3) (2016) 469–500, http://link.springer.com/article/10.1007/s10515-014-0173-z.

[202] M. Alshraideh, L. Bottaci, B.A. Mahafzah, Using program data-state scarcity to guide automatic test data generation, Softw. Qual. J. 18 (1) (2010) 109–144, http://link.springer.com/article/10.1007/s11219-009-9083-x.

[203] S. Krishnamoorthy, M.S. Hsiao, L. Lingappan, Strategies for scalable symbolic execution-driven test generation for programs, Science China Information Sciences 54 (9) (2011) 1797. ISSN: 1674-733X, 1869-1919, http://link.springer.com/article/10.1007/s11432-011-4368-7.

[204] K. Kähkönen, O. Saarikivi, K. Heljanko, Unfolding based automated testing of multithreaded programs, Automated Software Engineering 22 (4) (2015) 475–515, http://link.springer.com/article/10.1007/s10515-014-0150-6.

[205] N. Rungta, E.G. Mercer, W. Visser, Efficient testing of concurrent programs with abstraction-guided symbolic execution, in: Model Checking Software, Springer, Berlin, Heidelberg, 2009, ISBN: 978-3-642-02651-5, pp. 174–191, 978-3-642-02652-2, http://link.springer.com/chapter/10.1007/978-3-642-02652-2_16.

[206] F.S. Babamir, A. Hatamizadeh, S.M. Babamir, M. Dabbaghian, A. Norouzi, Application of genetic algorithm in automatic software testing, in: Networked Digital Technologies, Springer, Berlin, Heidelberg, 2010, ISBN: 978-3-642-14305-2, pp. 545–552, 978-3-642-14306-9, http://link.springer.com/chapter/10.1007/978-3-642-14306-9_54.

[207] M. Harman, P. McMinn, A theoretical & empirical analysis of evolutionary testing and hill climbing for structural test data generation. in: Proceedings of the 2007 International Symposium on Software Testing and Analysis, ACM, New York, NY, USA, 2007, ISBN: 978-1-59593-734-6, pp. 73–83, https://doi.org/10.1145/1273463.1273475.

[208] S.L. Rhys, S. Poulding, J.A. Clark, Using automated search to generate test data for Matlab. in: Proceedings of the 11th Annual Conference on Genetic and Evolutionary Computation, ACM, New York, NY, USA, 2009, ISBN: 978-1-60558-325-9, pp. 1697–1704, https://doi.org/10.1145/1569901.1570128.

[209] R. Chopra, S. Madan, Reusing black box test paths for white box testing of websites. in: 2013 3rd IEEE International Advance Computing Conference (IACC), February, 2013, pp. 1345–1350, https://doi.org/10.1109/IAdCC.2013.6514424.

[210] R. Kannavara, C.J. Havlicek, B. Chen, M.R. Tuttle, K. Cong, S. Ray, F. Xie, Challenges and opportunities with concolic testing. in: 2015 National Aerospace and Electronics Conference (NAECON), 2015, pp. 374–378, https://doi.org/10.1109/NAECON.2015.7443099.

[211] S.J. Galler, B.K. Aichernig, Survey on test data generation tools, Int. J. Softw. Tools Technol. Transfer 16 (6) (2014) 727–751, http://link.springer.com/article/10.1007/ s10009-013-0272-3.

[212] H. Liu, F.-C. Kuo, T.Y. Chen, Comparison of adaptive random testing and random testing under various testing and debugging scenarios, Softw. Practi. Exp. 42 (8) (2011) 1055–1074, http://onlinelibrary.wiley.com/doi/abs/10.1002/spe.1113.

[213] H. Liu, X. Xie, J. Yang, Y. Lu, T.Y. Chen, Adaptive random testing through test profiles, Softw. Practi. Exp. 41 (10) (2011) 1131–1154, http://onlinelibrary.wiley.com/ doi/abs/10.1002/spe.1067.

[214] K. Go, S. Kang, J. Baik, M. Kim, Pairwise testing for systems with data derived from real-valued variable inputs. Softw. Pract. Exp. 46 (3) (2014) 381–403, https://doi.org/ 10.1002/spe.2295.

[215] G. Fraser, N. Walkinshaw, Assessing and generating test sets in terms of behavioural adequacy, Softw. Testing Verification Reliab. 25 (8) (2015) 749–780, http:// onlinelibrary.wiley.com/doi/abs/10.1002/stvr.1575.

[216] L. Briand, Y. Labiche, Q. Lin, Improving the coverage criteria of UML state machines using data flow analysis, Softw. Testing Verification Reliab. 20 (3) (2009) 177–207, http://onlinelibrary.wiley.com/doi/abs/10.1002/stvr.410.

[217] Y. Liu, H. Zhu, An experimental evaluation of the reliability of adaptive random testing methods. in: 2008 Second International Conference on Secure System Integration and Reliability Improvement, July, 2008, pp. 24–31, https://doi.org/10.1109/SSIRI. 2008.18.

[218] S. Wu, Y. Wu, S. Xu, Acceleration of random testing for software. in: 2013 IEEE 19th Pacific Rim International Symposium on Dependable Computing, December, 2013, pp. 51–59, https://doi.org/10.1109/PRDC.2013.15.

[219] H. Liu, T.Y. Chen, Randomized quasi-random testing. IEEE Trans. Comput. 65 (6) (2016) 1896–1909. ISSN: 0018-9340, https://doi.org/10.1109/TC.2015.2455981.

[220] N. Walkinshaw, G. Fraser, Uncertainty-driven black-box test data generation. in: 2017 IEEE International Conference on Software Testing, Verification and Validation (ICST), March, 2017, pp. 253–263, https://doi.org/10.1109/ICST.2017.30.

[221] E.P. Enoiu, A. Cauevic, D. Sundmark, P. Pettersson, A controlled experiment in testing of safety-critical embedded software. in: 2016 IEEE International Conference on Software Testing, Verification and Validation (ICST), April, 2016, pp. 1–11, https:// doi.org/10.1109/ICST.2016.15.

[222] S.K. Khalsa, Y. Labiche, J. Nicoletta, The power of single and error annotations in category partition testing: an experimental evaluation. in: Proceedings of the 20th International Conference on Evaluation and Assessment in Software Engineering, ACM, New York, NY, USA, 2016, ISBN: 978-1-4503-3691-8, pp. 28:1–28:10, https://doi.org/10.1145/2915970.2915999.

[223] M.F. Granda, N. Condori-Fernández, T.E.J. Vos, O. Pastor, Effectiveness assessment of an early testing technique using model-level mutants. in: Proceedings of the 21st International Conference on Evaluation and Assessment in Software Engineering, ACM, New York, NY, USA, 2017, ISBN: 978-1-4503-4804-1, pp. 98–107, https://doi.org/10.1145/3084226.3084257.

[224] A. Arcuri, L. Briand, Adaptive random testing: an illusion of effectiveness? in: Proceedings of the 2011 International Symposium on Software Testing and Analysis, ACM, New York, NY, USA, 2011, ISBN: 978-1-4503-0562-4, pp. 265–275, https://doi.org/10.1145/2001420.2001452.

[225] D. Shin, E. Jee, D.-H. Bae, Comprehensive analysis of FBD test coverage criteria using mutants, Softw. Syst. Model. 15 (3) (2016) 631–645, http://link.springer.com/article/ 10.1007/s10270-014-0428-y.

[226] M. Ellims, D. Ince, M. Petre, The effectiveness of T-way test data generation, in: Computer Safety, Reliability, and Security, Springer, Berlin, Heidelberg, 2008,

ISBN: 978-3-540-87697-7, pp. 16–29, 978-3-540-87698-4, http://link.springer.com/chapter/10.1007/978-3-540-87698-4_5.

[227] D. Shin, E. Jee, D.-H. Bae, Empirical evaluation on FBD model-based test coverage criteria using mutation analysis, in: Model Driven Engineering Languages and Systems, Springer, Berlin, Heidelberg, 2012, ISBN: 978-3-642-33665-2, pp. 465–479, 978-3-642-33666-9, http://link.springer.com/chapter/10.1007/978-3-642-33666-9_30.

[228] S. Wang, J. Offutt, Comparison of unit-level automated test generation tools. in: 2009 International Conference on Software Testing, Verification, and Validation Workshops, 2009, pp. 210–219, https://doi.org/10.1109/ICSTW.2009.36.

[229] L.S. Ghandehari, J. Czerwonka, Y. Lei, S. Shafiee, R. Kacker, R. Kuhn, An empirical comparison of combinatorial and random testing. in: 2014 IEEE Seventh International Conference on Software Testing, Verification and Validation Workshops, 2014, pp. 68–77, https://doi.org/10.1109/ICSTW.2014.8.

[230] R.M.L.M. Moreira, A.C. Paiva, M. Nabuco, A. Memon, Pattern-based GUI testing: bridging the gap between design and quality assurance, Softw. Testing Verification Reliab. 27 (3) (2017) e1629. ISSN: 1099-1689, http://onlinelibrary.wiley.com/doi/abs/10.1002/stvr.1629.

[231] S.K. Khalsa, Y. Labiche, An extension of category partition testing for highly constrained systems. in: 2016 IEEE 17th International Symposium on High Assurance Systems Engineering (HASE), 2016, pp. 47–54, https://doi.org/10.1109/HASE.2016.45.

[232] M. Bures, B.S. Ahmed, On the effectiveness of combinatorial interaction testing: a case study. in: 2017 IEEE International Conference on Software Quality, Reliability and Security Companion (QRS-C), July, 2017, pp. 69–76, https://doi.org/10.1109/QRS-C.2017.20.

[233] H. Aljumaily, D. Cuadra, P. Martinez, Applying black-box testing to UML/OCL database models, Softw. Qual. J. 22 (2) (2014) 153–184, http://link.springer.com/article/10.1007/s11219-012-9192-9.

[234] E.F. de Souza, N.L. Vijaykumar, V.A. de Santiago Júnior, H-switch cover: a new test criterion to generate test case from finite state machines, Softw. Quality J. 25 (2) (2017) 373–405, http://link.springer.com/article/10.1007/s11219-015-9300-8.

[235] E.C.B de Matos, A.M. Moreira, J.B. de Souza Neto, An empirical study of test generation with BETA, J. Braz. Comput. Soc. 22 (1) (2016) 8, 1678-4804, http://link.springer.com/article/10.1186/s13173-016-0048-1.

[236] P. Li, T. Huynh, M. Reformat, J. Miller, A practical approach to testing GUI systems, Empir. Softw. Eng. 12 (4) (2007) 331–357, http://link.springer.com/article/10.1007/s10664-006-9031-3.

[237] B.K. Aichernig, J. Auer, E. Jobstl, R. Korosec, W. Krenn, R. Schlick, B.V. Schmidt, Model-based mutation testing of an industrial measurement device, in: Tests and Proofs, Springer, Cham, 2014, pp. 1–19, http://link.springer.com/chapter/10.1007/978-3-319-09099-3_1.

[238] G. Batra, J. Sengupta, An efficient metamorphic testing technique using genetic algorithm, in: Information Intelligence, Systems, Technology and Management, March, Springer, Berlin, Heidelberg, 2011, ISBN: 978-3-642-19422-1, pp. 180–188, 978-3-642-19423-8, http://link.springer.com/chapter/10.1007/978-3-642-19423-8_19.

[239] M. Prasanna, K.R. Chandran, Automated test case generation for object oriented systems using UML object diagrams, in: High Performance Architecture and Grid Computing, July, Springer, Berlin, Heidelberg, 2011, ISBN: 978-3-642-22576-5, pp. 417–423, 978-3-642-22577-2, http://link.springer.com/chapter/10.1007/978-3-642-22577-2_56.

[240] M.L.P. Souza, F.F. Silveira, A model-based testing method for dynamic aspect-oriented software, in: Computational Science and Its Applications—ICCSA 2017,

July, Springer, Cham, 2017, ISBN: 978-3-319-62406-8, pp. 95–111, 978-3-319-62407-5, http://link.springer.com/chapter/10.1007/978-3-319-62407-5_7.

[241] T.Y. Chen, F.C. Kuo, H. Liu, W.E. Wong, Code coverage of adaptive random testing. IEEE Transactions on Reliability 62 (1) (2013) 226–237, https://doi.org/10.1109/TR.2013.2240898.

[242] F. Hübner, W.-L. Huang, J. Peleska, Experimental evaluation of a novel equivalence class partition testing strategy, Softw. Syst. Model. (2017) 1–21, http://link.springer.com/article/10.1007/s10270-017-0595-8.

[243] F. Kurth, S. Schupp, S. Weißleder, Generating test data from a UML activity using the AMPL interface for constraint solvers, in: Tests and Proofs, Springer, Cham, 2014, ISBN: 978-3-319-09098-6, pp. 169–186, 978-3-319-09099-3, http://link.springer.com/chapter/10.1007/978-3-319-09099-3_14.

[244] L. Mariani, M. Pezzé, O. Riganelli, M. Santoro, Automatic testing of GUI-based applications, Softw. Testing Verification Reliab. 24 (5) (2014) 341–366. ISSN: 1099-1689, http://onlinelibrary.wiley.com/doi/abs/10.1002/stvr.1538.

[245] T. Yu, W. Srisa-an, G. Rothermel, An automated framework to support testing for process-level race conditions, Softw. Testing Verification Reliab. 27 (4-5) (2017) e1634. ISSN: 1099-1689, http://onlinelibrary.wiley.com/doi/abs/10.1002/stvr.1634.

[246] Y. Lei, R. Kacker, D.R. Kuhn, V. Okun, J. Lawrence, IPOG/IPOG-D: efficient test generation for multi-way combinatorial testing. Softw. Testing Verification Reliab. 18 (3) (2007) 125–148. ISSN: 1099-1689, https://doi.org/10.1002/stvr.381.

[247] Y. Lei, R.H. Carver, R. Kacker, D. Kung, A combinatorial testing strategy for concurrent programs, Softw. Testing Verification Reliab. 17 (4) (2007) 207–225. ISSN: 1099-1689, http://onlinelibrary.wiley.com/doi/abs/10.1002/stvr.369.

[248] M. Satpathy, A. Yeolekar, P. Peranandam, S. Ramesh, Efficient coverage of parallel and hierarchical stateflow models for test case generation, Softw. Testing Verification Reliab. 22 (7) (2011) 457–479. ISSN: 1099-1689, http://onlinelibrary.wiley.com/doi/abs/10.1002/stvr.444.

[249] Y. Lin, X. Tang, Y. Chen, J. Zhao, A divergence-oriented approach to adaptive random testing of java programs. in: 2009 IEEE/ACM International Conference on Automated Software Engineering, November, 2009, pp. 221–232, https://doi.org/10.1109/ASE.2009.13.

[250] M.E. Youmi, B. Falah, Testing web applications by unifying Fuzzy and All-Pairs techniques. in: 2014 International Conference on Multimedia Computing and Systems (ICMCS), April, 2014, pp. 547–551, https://doi.org/10.1109/ICMCS.2014.6911145.

[251] W.A. Ballance, S. Vilkomir, W. Jenkins, Effectiveness of pair-wise testing for software with Boolean inputs. in: Verification and Validation 2012 IEEE Fifth International Conference on Software Testing, Aplril, 2012, pp. 580–586, https://doi.org/10.1109/ICST.2012.144.

[252] C. Chow, T.Y. Chen, T.H. Tse, The ART of divide and conquer: an innovative approach to improving the efficiency of adaptive random testing. in: 2013 13th International Conference on Quality Software, July, 2013, pp. 268–275, https://doi.org/10.1109/QSIC.2013.19.

[253] D. Amalfitano, A.R. Fasolino, P. Tramontana, Rich internet application testing using execution trace data. in: 2010 Third International Conference on Software Testing, Verification, and Validation Workshops, April, 2010, pp. 274–283, https://doi.org/10.1109/ICSTW.2010.34.

[254] K. Liu, T. Wo, L. Cui, A fine-grained fault detection technique based on the virtual machine monitor. in: 2013 International Conference on Cloud Computing and Big Data, December, 2013, pp. 275–282, https://doi.org/10.1109/CLOUDCOM-ASIA.2013.18.

[255] C. Bertolini, G. Peres, M. d'Amorim, A. Mota, An empirical evaluation of automated black box testing techniques for crashing GUIs. in: 2009 International Conference on Software Testing Verification and Validation, April, 2009, pp. 21–30, https://doi.org/ 10.1109/ICST.2009.27.

[256] A. Yamada, A. Biere, C. Artho, T. Kitamura, E.H. Choi, Greedy combinatorial test case generation using unsatisfiable cores, in: 2016 31st IEEE/ACM International Conference on Automated Software Engineering (ASE), September, 2016, pp. 614–624.

[257] S. Park, S. Lu, Y. Zhou, CTrigger: exposing atomicity violation bugs from their hiding places. in: Proceedings of the 14th International Conference on Architectural Support for Programming Languages and Operating Systems, ACM, New York, NY, USA, 2009, ISBN: 978-1-60558-406-5, pp. 25–36, https://doi.org/10.1145/1508244. 1508249.

[258] Z. Lai, S.C. Cheung, W.K. Chan, Detecting atomic-set serializability violations in multithreaded programs through active randomized testing. in: Proceedings of the 32nd ACM/IEEE International Conference on Software Engineering - Volume 1, ACM, New York, NY, USA, 2010, ISBN: 978-1-60558-719-6, pp. 235–244, https://doi.org/10.1145/1806799.1806836.

[259] Z. Lai, S.C. Cheung, W.K. Chan, Inter-context control-flow and data-flow test adequacy criteria for nesC applications. in: Proceedings of the 16th ACM SIGSOFT International Symposium on Foundations of Software Engineering, ACM, New York, NY, USA, 2008, ISBN: 978-1-59593-995-1, pp. 94–104, https://doi.org/ 10.1145/1453101.1453115.

[260] A.F. Tappenden, J. Miller, Automated cookie collection testing. ACM Trans. Softw. Eng. Methodol. 23 (1) (2014) 3:1–3:40, https://doi.org/10.1145/2559936.

[261] K. Yatoh, K. Sakamoto, F. Ishikawa, S. Honiden, Feedback-controlled random test generation, in: Proceedings of the 2015 International Symposium on Software Testing and Analysis, ACM, New York, NY, USA, 2015, ISBN: 978-1-4503-3620-8, pp. 316–326, http://doi.acm.org/10.1145/2771783.2771805.

[262] H.Zhong, L. Zhang, S. Khurshid, Combinatorial generation of structurally complex test inputs for commercial software applications. in: Proceedings of the 2016 24th ACM SIGSOFT International Symposium on Foundations of Software Engineering, ACM, New York, NY, USA, 2016, ISBN: 978-1-4503-4218-6, pp. 981–986, https://doi. org/10.1145/2950290.2983959.

[263] A. Calvagna, A. Fornaia, E. Tramontana, Random versus combinatorial effectiveness in software conformance testing: a case study. in: Proceedings of the 30th Annual ACM Symposium on Applied Computing, ACM, New York, NY, USA, 2015, ISBN: 978-1-4503-3196-8, pp. 1797–1802, https://doi.org/10.1145/ 2695664.2695905.

[264] J.L. San Miguel, S. Takada, GUI and usage model-based test case generation for Android applications with change analysis. in: Proceedings of the 1st International Workshop on Mobile Development, ACM, New York, NY, USA, 2016, ISBN: 978-1-4503-4643-6, pp. 43–44, https://doi.org/10.1145/3001854.3001865.

[265] D. Amalfitano, A.R. Fasolino, P. Tramontana, S. De Carmine, A.M. Memon, Using GUI ripping for automated testing of Android applications. in: Proceedings of the 27th IEEE/ACM International Conference on Automated Software Engineering, ACM, New York, NY, USA, 2012, ISBN: 978-1-4503-1204-2, pp. 258–261, https://doi. org/10.1145/2351676.2351717.

[266] D. Marijan, A. Gotlieb, S. Sen, A. Hervieu, Practical Pairwise testing for software product lines. in: Proceedings of the 17th International Software Product Line Conference, ACM, New York, NY, USA, 2013, ISBN: 978-1-4503-1968-3, pp. 227–235, https://doi.org/10.1145/2491627.2491646.

[267] P.A. Brooks, A.M. Memon, Automated GUI testing guided by usage profiles. in: Proceedings of the Twenty-second IEEE/ACM International Conference on Automated Software Engineering, ACM, New York, NY, USA, 2007, ISBN: 978-1-59593-882-4, pp. 333–342, https://doi.org/10.1145/1321631.1321681.

[268] P. Thomson, A.F. Donaldson, A. Betts, Concurrency testing using schedule bounding: an empirical study. in: Proceedings of the 19th ACM SIGPLAN Symposium on Principles and Practice of Parallel Programming, ACM, New York, NY, USA, 2014, ISBN: 978-1-4503-2656-8, pp. 15–28, https://doi.org/10.1145/2555243.2555260.

[269] S.Apel, A. von Rhein, P. Wendler, A. Größlinger, D. Beyer, Strategies for product-line verification: case studies and experiments, in: Proceedings of the 2013 International Conference on Software Engineering, IEEE Press, Piscataway, NJ, USA, 2013, ISBN: 978-1-4673-3076-3, pp. 482–491, http://dl.acm.org/citation.cfm?id=2486788.2486852.

[270] C. Mao, T.Y. Chen, F.-C. Kuo, Out of sight, out of mind: a distance-aware forgetting strategy for adaptive random testing, Sci. China Inf. Sci. 60 (9) (2017) 092106. ISSN: 1674-733X, 1869-1919, http://link.springer.com/article/10.1007/s11432-016-0087-0.

[271] J.M. Balera, V.A. de Santiago Júnior, An algorithm for combinatorial interaction testing: definitions and rigorous evaluations. J. Softw. Eng. Res. Dev. 5 (1) (2017) 10, http://link.springer.com/article/10.1186/s40411-017-0043-z.

[272] X.-F. Qi, Z.-Y. Wang, J.-Q. Mao, P. Wang, Automated testing of web applications using combinatorial strategies, J. Comput. Sci. Technol. 32 (1) (2017) 199–210, http://link.springer.com/article/10.1007/s11390-017-1699-x.

[273] W. Zhang, B. Wei, H. Du, An output-oriented approach of test data generation based on genetic algorithm, in: Algorithms and Architectures for Parallel Processing, November, Springer, Cham, 2015, ISBN: 978-3-319-27160-6, pp. 100–108, 978-3-319-27161-3, http://link.springer.com/chapter/10.1007/978-3-319-27161-3_9.

[274] N. Walkinshaw, K. Bogdanov, J. Derrick, J. Paris, Increasing functional coverage by inductive testing: a case study, in: Testing software and systems, November, Springer, Berlin, Heidelberg, 2010, ISBN: 978-3-642-16572-6, pp. 126–141, 978-3-642-16573-3, http://link.springer.com/chapter/10.1007/978-3-642-16573-3_10.

[275] A. Gargantini, Using model checking to generate fault detecting tests, in: Tests and Proofs, Springer, Berlin, Heidelberg, 2007, ISBN: 978-3-540-73769-8, pp. 189–206, 978-3-540-73770-4, http://link.springer.com/chapter/10.1007/978-3-540-73770-4_11.

[276] S. Khatun, K.F. Rabbi, C.Y. Yaakub, M.F.J. Klaib, M.M. Ahmed, PS2Way: an efficient pairwise search approach for test data generation, in: Software Engineering and Computer Systems, Springer, Berlin, Heidelberg, 2011, ISBN: 978-3-642-22202-3, pp. 99–108, 978-3-642-22203-0, http://link.springer.com/chapter/10.1007/978-3-642-22203-0_9.

[277] M. Nabuco, A.C.R. Paiva, Model-based test case generation for Web applications, in: Computational Science and Its Applications–ICCSA 2014, July, Springer, Cham, 2014, ISBN: 978-3-319-09152-5, pp. 248–262, 978-3-319-09153-2, http://link.springer.com/chapter/10.1007/978-3-319-09153-2_19.

[278] R. Bloem, D. Hein, F. Röck, R. Schumi, Case study: automatic test case generation for a secure cache implementation, in: Tests and Proofs, Springer, Cham, 2015, ISBN: 978-3-319-21214-2, pp. 58–75, 978-3-319-21215-9, http://link.springer.com/chapter/10.1007/978-3-319-21215-9_4.

[279] C. Pacheco, S.K. Lahiri, M.D. Ernst, T. Ball, Feedback-directed random test generation. in: 29th International Conference on Software Engineering (ICSE'07), May, 2007, pp. 75–84, https://doi.org/10.1109/ICSE.2007.37.

[280] T.Y. Chen, F.C. Kuo, H. Liu, Distribution metric driven adaptive random testing. in: Seventh International Conference on Quality Software (QSIC 2007), October, 2007, pp. 274–279, https://doi.org/10.1109/QSIC.2007.4385507.

[281] A.F. Tappenden, J. Miller, A novel evolutionary approach for adaptive random testing. IEEE Trans. Reliab. 58 (4) (2009) 619–633. ISSN: 0018-9529, https://doi.org/10. 1109/TR.2009.2034288.

[282] R. Awedikian, B. Yannou, A practical model-based statistical approach for generating functional test cases: application in the automotive industry, Softw. Testing Verification Reliab. 24 (2) (2012) 85–123, http://onlinelibrary.wiley.com/doi/abs/ 10.1002/stvr.1479.

[283] S.E. Sprenkle, L.L. Pollock, L.M. Simko, Configuring effective navigation models and abstract test cases for web applications by analysing user behaviour. Softw. Testing Verification Reliab. 23 (6) (2013) 439–464, https://doi.org/10.1002/stvr.1496.

[284] H. Liu, T.Y. Chen, An innovative approach to randomising quasi-random sequences and its application into software testing. in: 2009 Ninth International Conference on Quality Software, August, 2009, pp. 59–64, https://doi.org/10.1109/QSIC.2009.16.

[285] S. Iftikhar, M.Z. Iqbal, M.U. Khan, W. Mahmood, An automated model based testing approach for platform games. in: 2015 ACM/IEEE 18th International Conference on Model Driven Engineering Languages and Systems (MODELS), 2015, pp. 426–435, https://doi.org/10.1109/MODELS.2015.7338274.

[286] S.M.B. Bhargavi, S.B. Nandeeswar, V. Suma, J.J. Rao, Conventional testing and combinatorial testing: a comparative analysis. in: 2016 International Conference on Inventive Computation Technologies (ICICT), Augustvol. 1, 2016, pp. 1–5, https://doi.org/10.1109/INVENTIVE.2016.7823200.

[287] S. Xu, L. Chen, C. Wang, O. Rud, A comparative study on black-box testing with open source applications. in: 2016 17th IEEE/ACIS International Conference on Software Engineering, Artificial Intelligence, Networking and Parallel/Distributed Computing (SNPD), May, 2016, pp. 527–532, https://doi.org/10.1109/SNPD. 2016.7515953.

[288] P. Wojciak, R. Tzoref-Brill, System level combinatorial testing in practice–The concurrent maintenance case study. in: Verification and Validation 2014 IEEE Seventh International Conference on Software Testing, March, 2014, pp. 103–112, https:// doi.org/10.1109/ICST.2014.23.

[289] L. Mariani, M. Pezzé, O. Riganelli, M. Santoro, AutoBlackTest: a tool for automatic black-box testing. in: 2011 33rd International Conference on Software Engineering (ICSE), May, 2011, pp. 1013–1015, https://doi.org/10.1145/1985793.1985979.

[290] L. Mei, W.K. Chan, T.H. Tse, Data flow testing of service choreography. in: Proceedings of the 7th Joint Meeting of the European Software Engineering Conference and the ACM SIGSOFT Symposium on The Foundations of Software Engineering, ACM, New York, NY, USA, 2009, ISBN: 978-1-60558-001-2, pp. 151–160, https://doi.org/10.1145/1595696.1595720.

[291] S. Bauersfeld, T.E.J. Vos, N. Condori-Fernandez, A. Bagnato, E. Brosse, Evaluating the TESTAR tool in an industrial case study. in: Proceedings of the 8th ACM/IEEE International Symposium on Empirical Software Engineering and Measurement, ACM, New York, NY, USA, 2014, ISBN: 978-1-4503-2774-9, pp. 4:1–4:9, https://doi.org/10.1145/2652524.2652588.

[292] R. Dixit, C. Lutteroth, G. Weber, FormTester: effective integration of model-based and manually specified test cases, in: Proceedings of the 37th International Conference on Software Engineering - Volume 2, IEEE Press, Piscataway, NJ, USA, 2015, pp. 745–748, http://dl.acm.org/citation.cfm?id=2819009.2819154.

[293] T. Azim, I. Neamtiu, Targeted and depth-first exploration for systematic testing of Android apps. in: Proceedings of the 2013 ACM SIGPLAN International Conference on Object Oriented Programming Systems Languages & Applications, ACM, New York, NY, USA, 2013, ISBN: 978-1-4503-2374-1, pp. 641–660, https://doi.org/ 10.1145/2509136.2509549.

[294] X. Devroey, G. Perrouin, P.-Y. Schobbens, Abstract test case generation for behavioural testing of software product lines. in: Proceedings of the 18th International Software Product Line Conference: Companion Volume for Workshops, Demonstrations and Tools - Volume 2, ACM, New York, NY, USA, 2014, ISBN: 978-1-4503-2739-8, pp. 86–93, https://doi.org/10.1145/2647908.2655971.

[295] K.-C. Chuang, C.-S. Shih, S.-H. Hung, User behavior augmented software testing for user-centered GUI. in: Proceedings of the 2011 ACM Symposium on Research in Applied Computation, ACM, New York, NY, USA, 2011, ISBN: 978-1-4503-1087-1, pp. 200–208, https://doi.org/10.1145/2103380.2103421.

[296] A.K. Jha, S. Lee, W.J. Lee, Modeling and test case generation of inter-component communication in Android, in: Proceedings of the Second ACM International Conference on Mobile Software Engineering and Systems, IEEE Press, Piscataway, NJ, USA, 2015, ISBN: 978-1-4799-1934-5, pp. 113–116, http://dl.acm.org/citation.cfm?id=2825041.2825061.

[297] J. Offutt, C. Alluri, An industrial study of applying input space partitioning to test financial calculation engines, Empir. Softw. Eng. 19 (3) (2014) 558–581. ISSN: 1382-3256, 1573-7616, http://link.springer.com/article/10.1007/s10664-012-9229-5.

[298] J. Offutt, V. Papadimitriou, U. Praphamontripong, A case study on bypass testing of web applications, Empir. Softw. Eng. 19 (1) (2014) 69–104. ISSN: 1382-3256, 1573-7616, http://link.springer.com/article/10.1007/s10664-012-9216-x.

[299] N. Razavi, A. Farzan, S.A. McIlraith, Generating effective tests for concurrent programs via AI automated planning techniques, Int. J. Softw. Tools Technol. Transfer 16 (1) (2014) 49–65. ISSN: 1433-2779, 1433-2787, http://link.springer.com/article/10.1007/s10009-013-0277-y.

[300] A.B. Sánchez, S. Segura, J.A. Parejo, A. Ruiz-Cortés, Variability testing in the wild: the Drupal case study, Softw. Syst. Model. 16 (1) (2017) 173–194. ISSN: 1619-1366, 1619-1374, http://link.springer.com/article/10.1007/s10270-015-0459-z.

[301] C. Wiederseiner, S.A. Jolly, V. Garousi, M.M. Eskandar, An open-source tool for automated generation of black-box xUnit test code and its industrial evaluation, in: Testing–Practice and Research Techniques, Springer, Berlin, Heidelberg, 2010, ISBN: 978-3-642-15584-0, pp. 118–128, 978-3-642-15585-7, http://link.springer.com/chapter/10.1007/978-3-642-15585-7_11.

[302] S. Arlt, P. Borromeo, M. Schäf, A. Podelski, Parameterized GUI tests, in: Testing Software and Systems, Springer, Berlin, Heidelberg, 2012, ISBN: 978-3-642-34690-3, pp. 247–262, 978-3-642-34691-0, http://link.springer.com/chapter/10.1007/978-3-642-34691-0_18.

[303] L. Frantzen, M.d.l.N. Huerta, Z.G. Kiss, T. Wallet, On-the-fly model-based testing of Web services with Jambition, in: Web Services and Formal Methods, Springer, Berlin, Heidelberg, 2008, ISBN: 978-3-642-01363-8, pp. 143–157, 978-3-642-01364-5, http://link.springer.com/chapter/10.1007/978-3-642-01364-5_9.

[304] M. Kläs, T. Bauer, A. Dereani, T. Söderqvist, P. Helle, A large-scale technology evaluation study: effects of model-based analysis and testing. in: 2015 IEEE/ACM 37th IEEE International Conference on Software Engineering2 2015, pp. 119–128, https://doi.org/10.1109/ICSE.2015.141. may vol.

[305] A. Mateen, K. Abbas, Optimization of model based functional test case generation for Android applications. in: 2017 IEEE International Conference on Power, Control, Signals and Instrumentation Engineering (ICPCSI), September, 2017, pp. 90–95, https://doi.org/10.1109/ICPCSI.2017.8391869.

[306] C. Tian, S. Liu, S. Nakajima, Utilizing model checking for automatic test case generation from conjunctions of predicates. in: 2011 IEEE Fourth International Conference on Software Testing, Verification and Validation Workshops, March, 2011, pp. 304–309, https://doi.org/10.1109/ICSTW.2011.45.

[307] M.U. Hayat, N. Qadeer, Intra component GUI test case generation technique. in: 2007 International Conference on Information and Emerging Technologies, July, 2007, pp. 1–5, https://doi.org/10.1109/ICIET.2007.4381328.

[308] H. Tang, G. Wu, J. Wei, H. Zhong, Generating test cases to expose concurrency bugs in Android applications. in: Proceedings of the 31st IEEE/ACM International Conference on Automated Software Engineering, ACM, New York, NY, USA, 2016, ISBN: 978-1-4503-3845-5, pp. 648–653, https://doi.org/10.1145/2970276. 2970320.

[309] R. Lefticaru, F. Ipate, An improved test generation approach from extended finite state machines using genetic algorithms, in: Software Engineering and Formal Methods, Springer, Berlin, Heidelberg, 2012, ISBN: 978-3-642-33825-0, pp. 293–307, 978-3-642-33826-7, http://link.springer.com/chapter/10.1007/978-3-642-33826-7_20.

[310] A. Arcuri, M.Z. Iqbal, L. Briand, Formal analysis of the effectiveness and predictability of random testing. in: Proceedings of the 19th International Symposium on Software Testing and Analysis, ACM, New York, NY, USA, 2010, ISBN: 978-1-60558-823-0, pp. 219–230, https://doi.org/10.1145/1831708.1831736.

[311] R.H. Carver, Y. Lei, Distributed reachability testing of concurrent programs, Concurrency and Computation: Practice and Experience 22 (18) (2010) 2445–2466, http://onlinelibrary.wiley.com/doi/abs/10.1002/cpe.1573.

[312] A. Calvagna, A. Gargantini, T-wise combinatorial interaction test suites construction based on coverage inheritance, Softw. Testing Verification Reliab. 22 (7) (2011) 507–526, http://onlinelibrary.wiley.com/doi/abs/10.1002/stvr.466.

[313] M.I. Younis, K.Z. Zamli, MC-MIPOG: a parallel t-way test generation strategy for multicore systems, ETRI J. 32 (1) (2010) 73–83, http://onlinelibrary.wiley.com/doi/abs/10.4218/etrij.10.0109.0266.

[314] R.C. Bryce, C.J. Colbourn, A density-based greedy algorithm for higher strength covering arrays, Softw. Testing Verification Reliab. 19 (1) (2008) 37–53, http://onlinelibrary.wiley.com/doi/abs/10.1002/stvr.393.

[315] A.T. Endo, A. Simao, Event tree algorithms to generate test sequences for composite Web services, Softw. Testing Verification Reliab. 29 (3) (2017) e1637, http://onlinelibrary.wiley.com/doi/abs/10.1002/stvr.1637.

[316] P. Accioly, P. Borba, R. Bonifácio, Comparing two black-box testing strategies for software product lines. in: 2012 Sixth Brazilian Symposium on Software Components, Architectures and Reuse, September, 2012, pp. 1–10, https://doi.org/10.1109/SBCARS.2012.17.

[317] L. Cordeiro, B. Fischer, Verifying multi-threaded software using SMT-based context-bounded model checking. in: 2011 33rd International Conference on Software Engineering (ICSE), May, 2011, pp. 331–340, https://doi.org/10.1145/1985793. 1985839.

[318] C. Nie, H. Wu, Y. Liang, H. Leung, F.C. Kuo, Z. Li, Search based combinatorial testing. in: 2012 19th Asia-Pacific Software Engineering Conference, December 1 2012, pp. 778–783, https://doi.org/10.1109/APSEC.2012.16. vol.

[319] S. Ali, M.Z. Iqbal, A. Arcuri, L.C. Briand, Generating test data from OCL constraints with search techniques. IEEE Trans. Softw. Eng. 39 (10) (2013) 1376–1402, https://doi.org/10.1109/TSE.2013.17.

[320] H. Hu, C.H. Jiang, F. Ye, K.Y. Cai, D. Huang, S.S. Yau, A parallel implementation strategy of adaptive testing. in: 2010 IEEE 34th Annual Computer Software and Applications Conference Workshops, 2010, pp. 214–219, https://doi.org/10.1109/COMPSACW.2010.44.

[321] T.B.N. Do, T. Kitamura, V.T. Nguyen, G. Hatayama, S. Sakuragi, H. Ohsaki, Constructing test cases for N-wise testing from tree-based test models. in: Proceedings of the Fourth Symposium on Information and Communication

Technology, ACM, New York, NY, USA, 2013, ISBN: 978-1-4503-2454-0, pp. 275–284, https://doi.org/10.1145/2542050.2542074.

[322] W. Xu, S. Kashyap, C. Min, T. Kim, Designing new operating primitives to improve fuzzing performance. in: Proceedings of the 2017 ACM SIGSAC Conference on Computer and Communications Security, ACM, New York, NY, USA, 2017, ISBN: 978-1-4503-4946-8, pp. 2313–2328, https://doi.org/10.1145/3133956.3134046.

[323] K. Sen, Effective random testing of concurrent programs. in: Proceedings of the Twenty-second IEEE/ACM International Conference on Automated Software Engineering, ACM, New York, NY, USA, 2007, ISBN: 978-1-59593-882-4, pp. 323–332, https://doi.org/10.1145/1321631.1321679.

[324] A. Marchetto, P. Tonella, Using search-based algorithms for Ajax event sequence generation during testing, Empir. Softw. Eng. 16 (1) (2011) 103–140, http://link.springer.com/article/10.1007/s10664-010-9149-1.

[325] R.-Z. Qi, Z.-J. Wang, S.-Y. Li, A parallel genetic algorithm based on spark for pairwise test suite generation, J. Comput. Sci. Technol. 31 (2) (2016) 417–427, http://link.springer.com/article/10.1007/s11390-016-1635-5.

[326] B.K. Aichernig, D. Ničković, S. Tiran, Scalable incremental test-case generation from large behavior models, in: Tests and Proofs, Springer, Cham, 2015, ISBN: 978-3-319-21214-2, pp. 1–18, 978-3-319-21215-9, http://link.springer.com/chapter/10.1007/978-3-319-21215-9_1.

[327] K. Rabbi, Q. Mamun, An effective t-way test data generation strategy, in: Security and Privacy in Communication Networks, Springer, Cham, 2015, ISBN: 978-3-319-28864-2, pp. 633–648, 978-3-319-28865-9, http://link.springer.com/chapter/10.1007/978-3-319-28865-9_42.

[328] M. Shahbaz, R. Groz, Analysis and testing of black-box component-based systems by inferring partial models, Softw. Testing Verification Reliab. 24 (4) (2013) 253–288. ISSN: 1099-1689, http://onlinelibrary.wiley.com/doi/abs/10.1002/stvr.1491.

[329] E. Borjesson, Industrial applicability of visual GUI testing for system and acceptance test automation. in: Verification and Validation 2012 IEEE Fifth International Conference on Software Testing, April, 2012, pp. 475–478, https://doi.org/10.1109/ICST.2012.129.

[330] R. Gao, J.S. Eo, W.E. Wong, X. Gao, S.-Y. Lee, An empirical study of requirements-based test generation on an automobile control system. in: Proceedings of the 29th Annual ACM Symposium on Applied Computing, ACM, New York, NY, USA, 2014, ISBN: 978-1-4503-2469-4, pp. 1094–1099, https://doi.org/10.1145/2554850.2554934.

[331] M.Z. Iqbal, A. Arcuri, L. Briand, Environment modeling and simulation for automated testing of soft real-time embedded software, Software & Systems Modeling 14 (1) (2015) 483–524, http://link.springer.com/article/10.1007/s10270-013-0328-6.

[332] R.S. Zybin, V.V. Kuliamin, A.V. Ponomarenko, V.V. Rubanov, E.S. Chernov, Automation of broad sanity test generation, Programming and Computer Software 34 (6) (2008) 351–363, http://link.springer.com/article/10.1134/S0361768808060066.

[333] A. Bhat, S.M.K. Quadri, Equivalence class partitioning and boundary value analysis - a review, in: 2015 2nd International Conference on Computing for Sustainable Global Development (INDIACom), March, 2015, pp. 1557–1562.

[334] E.P. Enoiu, A. Čaušević, T.J. Ostrand, E.J. Weyuker, D. Sundmark, P. Pettersson, Automated test generation using model checking: an industrial evaluation, Int. J. Softw. Tools Technol. Transfer 18 (3) (2016) 335–353, http://link.springer.com/article/10.1007/s10009-014-0355-9.

[335] A. Arcuri, M.Z. Iqbal, L. Briand, Black-box system testing of real-time embedded systems using random and search-based testing, in: Testing Software and Systems, Springer, Berlin, Heidelberg, 2010, ISBN: 978-3-642-16572-6, pp. 95–110, 978-3-642-16573-3, http://link.springer.com/chapter/10.1007/978-3-642-16573-3_8.

[336] H.M. Sneed, S. Huang, The design and use of WSDL-Test: a tool for testing Web services, J. Softw. Maint. Evol. Res. Pract. 19 (5) (2007) 297–314. ISSN: 1532-0618, http://onlinelibrary.wiley.com/doi/abs/10.1002/smr.354.

[337] E. Borjesson, R. Feldt, Automated System testing using visual GUI testing tools: a comparative study in industry. in: Verification and Validation 2012 IEEE Fifth International Conference on Software Testing, April, 2012, pp. 350–359, https://doi.org/10.1109/ICST.2012.115.

[338] C. Schwarzl, B. Peischl, Generation of executable test cases based on behavioral UML system models. in: Proceedings of the 5th Workshop on Automation of Software Test, ACM, New York, NY, USA, 2010, ISBN: 978-1-60558-970-1, pp. 31–34, https://doi.org/10.1145/1808266.1808271.

[339] E. Puoskari, T.E.J. Vos, N. Condori-Fernandez, P.M. Kruse, Evaluating applicability of combinatorial testing in an industrial environment: a case study. in: Proceedings of the 2013 International Workshop on Joining AcadeMiA and Industry Contributions to Testing Automation, ACM, New York, NY, USA, 2013, ISBN: 978-1-4503-2161-7, pp. 7–12, https://doi.org/10.1145/2489280.2489287.

[340] T.E.J. Vos, F.F. Lindlar, B. Wilmes, A. Windisch, A.I. Baars, P.M. Kruse, H. Gross, J. Wegener, Evolutionary functional black-box testing in an industrial setting, Softw. Qual. J. 21 (2) (2013) 259–288, http://link.springer.com/article/10.1007/s11219-012-9174-y.

[341] A.M. Memon, An event-flow model of GUI-based applications for testing, Softw. Testing Verification Reliab. 17 (3) (2007) 137–157, http://onlinelibrary.wiley.com/doi/abs/10.1002/stvr.364.

[342] F. Naseer, S.u. Rehman, K. Hussain, Using meta-data technique for component based black box testing. in: 2010 6th International Conference on Emerging Technologies (ICET), October, 2010, pp. 276–281, https://doi.org/10.1109/ICET.2010.5638474.

[343] M.A. Khan, M. Sadiq, Analysis of black box software testing techniques: a case study. in: The 2011 International Conference and Workshop on Current Trends in Information Technology (CTIT 11), 2011, pp. 1–5, https://doi.org/10.1109/CTIT.2011.6107931.

[344] W. Zheng, G. Bundell, Model-based software component testing: a UML-based approach. in: 6th IEEE/ACIS International Conference on Computer and Information Science (ICIS 2007), July, 2007, pp. 891–899, https://doi.org/10.1109/ICIS.2007.136.

[345] H.M. Rusli, M. Puteh, S. Ibrahim, S.G.H. Tabatabaei, A comparative evaluation of state-of-the-art web service composition testing approaches. in: Proceedings of the 6th International Workshop on Automation of Software Test, ACM, New York, NY, USA, 2011, ISBN: 978-1-4503-0592-1, pp. 29–35, https://doi.org/10.1145/1982595.1982602.

[346] L.C. Briand, A critical analysis of empirical research in software testing. in: First International Symposium on Empirical Software Engineering and Measurement (ESEM 2007), September, 2007, pp. 1–8, https://doi.org/10.1109/ESEM.2007.40.

About the authors

Mitchell Mayeda received his B.S. degree in Computer Science from the University of Denver, Denver, CO, United States, in 2015 and received a M.S. degree in Computer Science from the same university in 2019. He currently develops large scale microservices as a software engineer at SAP.

Dr. Anneliese Amschler Andrews received her PhD in Computer Science from Duke University. She has published over 200 articles in software engineering and performance modeling. She is Full Professor of Computer Science at the University of Denver. Her research interests include software testing, software reliability, software maintenance, and experimental software engineering. She has served on a number of editorial Boards, including as Editor-in-Chief of the IEEE Transactions on Software Engineering.

CHAPTER THREE

The screening phase in systematic reviews: Can we speed up the process?

Igor Rožanc[a] and Marjan Mernik[b]
[a]University of Ljubljana, Faculty of Computer and Information Science, Ljubljana, Slovenia
[b]University of Maribor, Faculty of Electrical Engineering and Computer Science, Maribor, Slovenia

Contents

Abstract

The aim of a systematic reviews (SRs) is to gain a better understanding of a certain aspect of selected research field using the principle of classification of a large number of carefully selected articles. Selection of a proper set of articles is a crucial yet delicate task, which demands a large portion of tedious manual work. This article proposes to automate the screening of a large set of articles while conducting an SR. A rigorous approach is described, which conforms with the SR guidelines, and a tool to efficiently support such an approach is presented as well. The effect of approach is presented by a demonstration experiment which compares its results with the results of a classic manual screening. Finally, the recommendations for the proper use of the approach (i.e., the size of the pilot set and decision rule structure) are presented.

1. Introduction

1.1 Motivation

The first phase of research work in any field is to get insight into the state of the art by reviewing the existing knowledge [1]. Diverse types of evidence-based research exist and one of them is systematic review (SR). A specific form of SR, namely systematic literature review (SLR) is performed over a large number of carefully selected articles in a field to achieve this aim [2]. An alternative is to conduct a systematic mapping study (SMS), which only classifies studies into a set of predefined clusters. As such it tends to give less specific answers but involves an even larger set of selected articles [3]. Both types of reviews have to follow a strictly defined protocol to be valid. This protocol is suitable when it is consistent with the guidelines for performing SR [4]. One of the most important steps is performed after the collection of a large number of approximately suitable articles from different sources when a set of suitable articles is selected by screening using defined inclusion/exclusion criteria. The screening is usually performed by a quick manual reading of each article's title and abstract [5, 6]. Then, the article is included or excluded from the set of suitable articles. Unfortunately, the amount of collected articles is usually very large (more than 5000) and demands a lot of effort [7]. Syriani reportedz that screening of 11,173 articles in total demanded more than 558 h [8], while Kosar screened 1636 articles in 81 h [9]. Even in case of a smaller initial set of articles, the manual screening consumes a large amount of effort (time) [10]. As a consequence, the SR performers tend to diminish the initial set of articles by limiting the initial search or performing it in a fast, not enough detailed way. This practice may lead to exclusion of valuable evidence. By suitable automation of screening several important advantages can be achieved [11, 12]. Most importantly, the shorter performance time with less manual effort enables the researcher to actually screen a larger set of collected articles. The difficult part is how to efficiently perform such an automated screening in a correct, yet appropriate way. Additional issues involve the usability concerns (e.g., the availability of a tool).

1.2 Objectives

The main objective of this work is to propose an approach for a faster, but still affordable screening of a large number of articles, which is compatible with the strict rules of SR protocol. The approach is supported by a tool, which can be configured to better meet the expected level of screening results correctness. It enables adaptable configuration of screening, iterative optimization of screening by using it on small pilot sets, screening execution and the assessment of the outcomes. Additionally, an experiment to demonstrate the suitability of such an approach is presented. The experiment compares the effect of the manual and several executions of the automated screening with an emphasis on the size of the pilot set and on the definition of decision rules. It is conducted on the collection of articles for an SMS on examining different processes while developing a domain-specific language (DSL) or domain-specific modeling language (DSML). Finally, the concluding remarks concisely points out the main benefits of the tool usage.

1.3 Research method

The proper management of an SLR or SMS must follow a rigorous protocol, thus a detailed insight into various aspects of both types of research are given first. Next, the problem of the most time-consuming parts of SR process is explored to find out possible points where time could be saved. Focusing on the step of the article's screening, a solution for the automation of this step is proposed. The solution is configurable and it is based on the principle of the statistical analysis of the article's contents. The automated screening approach is implemented by a tool. To demonstrate its value, an experiment is performed, too. The actual context of the experiment (i.e., the SMS on examining the different processes while developing DS(M)L) is explained first. The experiment compares the results of the manual vs automated screening execution. Several variants of tool's improvement strategies are explored in the experiment discussing four issues: the necessary size of the pilot set to be manually screened, the improvement strategy, the number and type of keywords, and the structure of decision rules. The presentation of results is accompanied by a short discussion. Finally, the related work is presented to compare our approach with similar solutions.

1.4 Expected results

First, the definition of an approach for the automated screening demonstrates its compatibility with SR protocol. Second, the custom-made tool we developed to support the approach proves its practical applicability. The tool is configurable and enables different levels of classification accuracy. Third, the results of an experiment demonstrate the practical level up to which screening step can be automated, and the required manual effort. The final part of the experiment shortly points out the main achievements of its use.

In the next section, the principles of the evidence-based research with the emphasis on the details of SR (SLR and SMS) are explained. In Section 3, the problem of screening automation and its configurable solution is presented in detail. Especially, an automatic screening approach is explained. In Section 4, the presentation of solution's implementation with a custom-made tool is explained. Section 5 describe a demonstration experiment we performed. Its context is given in the first part, while the core of this section presents four parts of the experiment with the discussion of corresponding results. The related work is presented in Section 6. The last section concludes and presents future work.

2. Evidence-based research in software engineering

The results of research work can contribute to several types of achievements [13], but they must clearly base on a solid foundation of previous experience. The term *evidence-based research* is implicated from *evidence-based practice (EBP)* in the field of medicine, which defines an up-to-date research evidence to inform clinical decisions [14]. In the field of medicine, this approach is enforced by a large corpus of predominantly well structured (experience and research) work. Unfortunately, this is not the case in many other research areas.

The research work in software engineering (SE) is mostly aimed at the presentation of the evidence, which is very versatile and less consistent with predefined rules [4]. This is due to the use of diverse empirical research methods in SE. The goal of this section is to provide insight into the empirical research methods in SE with emphasis on two types of SRs, namely SLRs and SMSs.

2.1 Empirical research methods

Wide view of empirical research methods in SE is given in [15]. Using previous research in this area, Zelkowitz and Wallace defined 12 types of research methods, which are divided into three groups:

(a) *Observational methods* collect relevant data during the development of a project and use them for further analysis.
- *Project Monitoring* can be defined as a collection of data the project development generates without a specific aim, and their use for some immediate analysis.
- *A case study* collects the data project generates over time as well, but the project is (slightly) modified to gather specific data to analyze a specific view of a project.
- *An assertion* is a demonstration of the effectiveness of a technology, which is used in an experiment performed by the developer.
- *A field study* gathers suitable data from many projects in the studied field and analyzes them to answer specific questions.

(b) *Historical methods* analyze the data from the already finished projects.
- *A literature search* analyzes the results of papers and other publicly available documents to provide information across a broad range of industries at a low cost.
- *Legacy data* is a low-cost form of experimentation, where the legacy data of finished project (e.g., source code, specifications, test results, quantitative data on artifacts) is used to study a specific objective.
- *Lessons learned* method is performed after a project is finished. The available data and other sources are qualitatively analyzed with the aim to improve the further development of similar projects.
- *A static analysis* checks the structure and views of a finished product to determine its specific characteristics.

(c) *Controlled methods* are classical research methods, which contain multiple instances of an observation in order to provide for the statistical validity of the results.
- *A replicated experiment* is a type of controlled method, which contains several parts (e.g., projects, subjects) performed in a real (industrial)

setting. Each part experiments one alternative solution, enabling a comparison of many or (in the ideal case) all possible solutions.

- *A synthetic experiment* is a cheaper version of the replicated experiment, which is performed in a smaller artificial environment. Consequently, its objectives are often smaller and its validity limited.
- *A dynamic analysis* is used to perform an evaluation of a product by its modification or execution under carefully controlled situations.
- *A simulation* is an evaluation of a technology by executing the product using a model of the real environment to predict, how the real environment will react to the new technology.

There are alternative definitions of research methods, which emphasize other aspects. Wieringa [16] proposed the following division of the research methods from the view of intended use:

- *An expert opinion* is the opinion of experts about the possible usability and usefulness of the artifact, which is obtained before the artifact is actually analyzed by researchers.
- *A single-case mechanism experiment* investigates one object in order to test the effect of some mechanism that the researcher believes to be present in it.
- *A technical action research* is a case-based mechanism experiment too, but it is also a real-world consultancy project. Here the researcher uses the experience to learn about the effects (in the uncontrolled conditions) in practice.
- *A statistical difference-making experiment* is an experiment performed on a relevant sample of the population to examine a statistical property of the population.
- *An observational case study* investigates a new technology in an industrial setting or an observational study of a project. It includes case and field studies.
- *A meta-research method* is usually a systematic literature review analysis of the experience on the specific research topic.
- *Methods to collect data* include different approaches aimed to collect and describe data without further analysis.
- *Techniques to infer information from data* have the emphasis on (the different approaches of) the analysis of data.

Another division of the empirical evidence is given in work of Kitchenham [4], who divides studies into *the primary*, *the secondary* and *the tertiary* studies.

- *The primary study* is an empirical study, which investigates a specific research question. Most of the research experience belong to this category.

- *The secondary study* aims to review all primary studies, which are related to a specific research question. It integrates all evidence related to a specific research question and provides an understanding of the research question's past experience.
- *The tertiary study* is a review of secondary studies related to the same research question.

The most of above presented empirical research methods produce a primary study. Our work is focused on secondary (and tertiary) studies, which are covered under *literature review/analysis* in [15] and *meta-research method* in [16].

2.2 Systematic literature review

Systematic literature review (SLR) is a *secondary* study that uses a well-defined methodology to identify, analyze and interpret all available evidence related to a specific research question in a way that is unbiased and repeatable to a certain degree [4]. Sometimes other names are used as well, namely review, systematic review, literature review or survey.

A rigorous procedure for performing SLR is a must to obtain trustworthy results. Eventually, strict SLR guidelines for reviewing evidence in SE were developed, which define each step of the review process in detail. The base for SLR guidelines was taken from the field of medicine [17], where strict formal rules for conducting and documenting research work are enforced for years.

Several authors were interested in defining a proper form of SLR research in SE [6, 18], most notably Barbara Kitchenham and her research group. They proposed several preliminary works toward the guidelines for conducting empirical research [1, 2, 19]. In 2007, she coauthored the official guidelines [4].

The main phases of each SLR process are

1. Planning the review
2. Conducting the review
3. Reporting the review

In Sections 2.2.1–2.2.3, the phases (and the steps they contain) are presented in detail. The steps are presented in a sequential way as they are usually conducted, but some of them can be performed in parallel as well. Some steps can be conducted in an iterative way, too.

2.2.1 Planning the review

In the first phase, the need for a review is identified. The review is planned by specifying research questions and defining a review protocol. Main steps of this phase are:

1. *Identification of the need for a review* is the entry step into any SLR, where the aim of the review, the sources, the restrictions, the inclusion/exclusion criteria, the performance of SLR steps and their quality are questioned. To find out those perspectives, authors in [20] proposed a number of specific questions on (a) the review's objectives, (b) the sources to investigate, (c) the inclusion/exclusion criteria, (d) the quality criteria, (e) the principles of data extraction and composition, and (f) the method applied to define conclusions from evidence.

   On the other hand, the Centre for Reviews and Dissemination of the University of York[a] summarized those aspects in four questions only:
   (a) Are the review's inclusion and exclusion criteria described and appropriate?
   (b) Is the literature search likely to cover all relevant studies?
   (c) Did the reviewers assess the quality/validity of the included studies?
   (d) Were the basic data/studies adequately described?
2. *Commissioning a review* step is not required. It is needed when the SLR is outsourced due to lack of expertise or time. A commissioning document has to be created to define the work to be performed. The most important parts describe the review questions, the expected methods of the review, the project timetable and the budget.
3. *Specifying the research question(s)* is a crucial part of any SLR as they actually drive the entire systematic review process. The research questions are used to identify articles included in SLR, how to extract the data (to address the research questions), and the way the conclusion summarizes their answers.

   The definition of suitable research questions is directed toward the right content and form. First, the research question in SE should fall into one of the following categories:
   * It assesses the effect of a software engineering technology.
   * It assesses the frequency (or rate) of a project development factor such as the adoption of a technology, or the frequency or rate of project success/failure.
   * It identifies the cost and risk factors associated with the technology.
   * It identifies the impact of the technologies on reliability, performance and cost models.
   * It performs a cost–benefit analysis of employing specific software development technologies or software applications.

---

[a] http://www.york.ac.uk/inst/crd/crddatabases.htm.

Second, the right question must be meaningful and important to the practitioners as well as the researchers. Finally, it should lead to changes in current software engineering practice or to increased confidence in the value of current practice. The question may identify discrepancies between commonly held beliefs and reality as well.

The structure of the research question enables the definition of the population addressed, the intervention performed and the outcome obtained. An extended framework for the question elements is defined as a PICOC (Population, Intervention, Comparison, Outcome, Context) criteria [21]:

- *Population* might address a specific software engineering role, a category of the software engineer, an application area, or an industry group.
- *Intervention* is the software procedure, technology or tool used to address a specific issue.
- *Comparison* defines the software engineering methodology, tool, technology, or procedure with which the intervention is being compared.
- *Outcomes* should be clearly presented from the view on how they relate to improved quality for practitioners.
- *Context* describes the context in which the comparison takes place (e.g., academia or industry), the actual participants included (e.g., practitioners, academics, students), and the scale of the tasks performed (e.g., small scale, large scale).

4. *Developing a review protocol* step specifies the methods that will be used to perform a specific systematic review and presents them as a defined protocol. The protocol usually includes the following elements:
   - *Background* is a clearly described rationale for the survey.
   - *The research questions* define what exactly the review intends to answer.
   - *The strategy* describe the search for primary studies including search terms and resources to be searched. Resources include digital libraries, specific journals, conference proceedings, and other sources.
   - *Study selection criteria* determines which studies are included or excluded from SR. A pilot study on a subset of primary studies helps to define the selection criteria.
   - *Study selection procedures* describe how the selection criteria will be applied, the number of SR performers, and how the disagreements will be resolved.

- *Study quality assessment checklists and procedures* define the formal quality check and they should be developed by the researchers to assess the quality of SR.
- *Data extraction strategy* defines how the information required for each primary study will be obtained.
- *Synthesis of the extracted data* specifies the (formal or nonformal) synthesis strategy, which is used in SLR.
- *Dissemination strategy* defines the presentation of the results.
- *Project timetable* defines the review schedule.
5. *Evaluating the review protocol* validates the correctness of the protocol by providing answers to questions like: are the search strings correctly derived from the research questions, do the extracted data properly address the research question(s), or whether the data analysis procedure appropriately answers the research questions.

2.2.2 Conducting the review

In the second phase, we actually perform the systematic review: the empirical evidence (articles) are collected, the suitable part of them is selected by screening and the data are extracted. The main steps are:

1. *Identification of research* defines a rigorous and unbiased search strategy to find as many as possible primary studies relating to the research question. It should be iterative, developed in cooperation with experts and benefit from potentially relevant studies (e.g., preliminary or trial searches). A general approach is to break down the question using the PICOC elements, define a list of keywords (e.g., synonyms, abbreviations, and alternative spellings) and construct search strings as a logical expression.

    The searches for primary studies can be undertaken in several ways.
    - Initially, the search is focused on digital libraries (e.g., IEEExplore, ACM Digital library, Web of Science, Google Scholar, Citeseer library, Inspec, ScienceDirect, EI Compendex, SpringerLink, SCOPUS, and others).
    - Later, other sources of evidence must also be searched (sometimes manually) including reference lists from relevant primary studies and reviewed articles (e.g., backward and forward snowballing [22]), journals and conference proceedings, gray literature (e.g., technical and other reports), research registers, and the Internet.
    - Finally, it is important to identify specific researchers to approach directly for advice on the appropriate source material.

Publication bias (i.e., the positive results are more likely to be published than negative ones) may lead to systematic bias in SR, too. The appropriate approach is to use alternative sources such as gray literature or directly contacting experts and researchers working in the area.

Bibliography management and document retrieval tools can assist the performer dealing with managing a large number of references obtained from the literature search. Once reference lists are finalized, the full articles of potentially useful studies have to be obtained.

To prove that the SLR process is performed in a transparent and repetitive way the SR must be documented in sufficient details. The search should be documented as it occurs including all the changes, and the unfiltered search results should be saved for later verification. Enough detail has to be documented (e.g., the name, the search strategy, date of search and years covered in case of DL search) as well as the rationale behind it.

2. *Selection of primary studies* assesses the actual relevance of all potentially relevant primary studies collected.

Study selection criteria identify the primary studies with direct evidence on the research question. To reduce the likelihood of bias, selection criteria should be defined in the protocol definition, but they may be refined during the search process. Inclusion and exclusion criteria are based on the research question. They should use pilot studies to be reliably interpreted and that they provide correct classification.

Study selection is performed in following steps.

- Initially, the screening is performed. The selection criteria are loose here and the papers are excluded based on title and abstract. For all included articles a full copy should be obtained. Brereton [6] points out that due to the poor quality of abstracts in SE the conclusions might be reviewed, too.
- Next, the inclusion/exclusion criteria based on practical issues are applied such as language, journal, authors, setting, participants or subjects, research design, sampling method, and date of publication.
- Sometimes, a more detailed third stage might be performed in the selection process.

A list of excluded studies with the reasons for exclusion should be maintained. As there might be a large number of totally irrelevant papers, a list might be formatted only after the totally irrelevant papers have been excluded.

If two or more researchers assess each paper, the agreement between researchers should be documented. Each disagreement must be discussed and resolved, the same as each uncertainty about the inclusion/exclusion of some studies. For this reason, expert's and senior researcher's opinions should be considered. They might apply a test–retest approach, and re-evaluate a random sample of the primary studies to check the consistency of decisions.

3. *Study quality assessment* is applied to provide a more detailed inclusion/exclusion criteria, to further investigate the quality of differences explanation, to weight the importance of individual studies, and to guide the interpretation of findings and further research. The quality relates to the extent to which the study minimizes bias (i.e., systematic error) and maximizes internal and external validity.

    *   An initial quality evaluation can be based on the applied type of experiment, so the performers might restrict the type of studies they include in their SLR.

    *   Detailed quality assessments are usually based on the quality checklists of needed factors to evaluate studies. Checklists are derived from factors that could bias the study results and they may have numerical scales. Usually, the selection, performance, measurement and exclusion bias should be considered [4].

    The factors identified might include generic (i.e., related to the features of particular study designs) and specific items (i.e., related to the review's subject area). Additionally, checklists should address the bias and validity problems in the different stages of SLR (e.g., design, conduct or analysis). A number of published quality checklists for different types of empirical study exist (e.g., in [21]).

    The quality data can be used to assist primary study selection (i.e., to construct detailed inclusion/exclusion criteria) or the data analysis and synthesis (i.e., to investigate the association of the quality differences with primary study outcomes). The quality data can be collected during the main data extraction using a joint form. In case of poorly reported primary studies, the reporters should attempt to obtain more information on quality from the authors of the study.

4. *Data extraction and monitoring* has the objective to design data extraction forms to accurately record the information. The data extraction forms should be defined and piloted when the study protocol is defined.

    *   The data extraction forms must be designed to collect all the information needed to address the review questions and the study quality

criteria which can require separate forms. Usually, data extraction will define a set of numerical values to be extracted for each study. The forms should include also standard information like the name of the reviewer, date of extraction, publication details, etc.

- The data extraction forms need to be piloted on a sample of primary studies by all the researchers involved. The pilot studies aim to assess the completeness and usability of the forms.

- The data extraction should be performed independently by two or more researchers, the outcomes must be compared and all disagreements resolved. When the agreement cannot be achieved, a sensitivity analysis should be performed. A separate form must be used to mark and correct errors or disagreements. When the researchers review different primary studies, they must use the same method in a consistent manner. If even this is not applicable, other checking techniques must be used (e.g., senior researchers perform data extraction on a random sample and check its consistency, or perform a test–retest check).

- Additionally, multiple publications of the same data should be excluded from the SR, and the better one should be included. As primary studies sometimes do not provide enough data, the manipulation of the studies in progress or other published data might provide missing information. In this case, the authors should be contacted, too.

5. *Data synthesis* involves the aggregation of the results of the included primary studies. Synthesis can be descriptive or quantitative. The first presents the extracted information about the studies in a clear and structured tabulated form. The results can be consistent or inconsistent, and they should display the impact of potential sources of this. Quantitative data should also be presented in tabular form including the statistical information (e.g., the sample size for each intervention, the difference between the mean values for each intervention, the confidence interval for the difference, etc.). As the quantitative synthesis results from different studies, the study outcomes are presented in different ways. A meta-analysis is a form of synthesis, where statistical techniques are used to obtain quantitative results. The data synthesis activities should be specified in the review protocol.

The means and variances of the quantitative results are usually presented by a forest plot. To avoid the problems of the post analysis, the possible sources of the difference (e.g., studies of a different type) should be identified when constructing the review protocol. Synthesizing qualitative studies involves an integration of studies using natural language results and conclusions. Synthesis of qualitative and quantitative studies should be synthesized separately or it should integrate the qualitative findings to explain the quantitative results.

Sensitivity analysis is important to determine whether the results are robust enough. It is usually performed as part of the meta-analysis. When a formal meta-analysis is not performed, the forest plots can identify high-quality primary studies. Publication bias may be presented using funnel plots.

2.2.3 Reporting the review

Finally, the report containing the findings has to be written and publicized. This is done in the third phase, which contains three steps:

1. *Specifying dissemination mechanisms* serves the purpose of efficient presentation of the results of an SR. It should be planned in advance (i.e., when preparing review protocol). The usual way is to publish them in academic journals and/or conferences, but other possibilities (e.g., nonacademic journals and magazines, posters, web pages) should be considered, too.
2. *Formatting the main report* should include (at least) one of the following formats: a technical report or (one section in) a Ph.D. thesis. When published as a journal or conference paper, the size restrictions limit the details, so a reference to a technical report or thesis is needed. The structure and contents of the report are adapted to the type of the publication.
3. *Evaluating the report* is the final step. Scientific journal articles or Ph.D. theses are rigidly reviewed, but also for other forms of dissemination a peer review should be performed by researchers with the expertise in the specific domain. The evaluation process can use the quality checklists for SLRs, too.

2.3 Systematic mapping study

Systematic mapping study (SMS) is a (broad) review of primary studies in a specific topic area that applies the principle of clustering to identify what evidence is available on the topic [4]. The results of an SMS may identify research areas for conducting an SLR. Sometimes, the name *scoping study* is used instead.

The main differences between SLRs and SMSs concern the principle and the goal of the research, but influence its execution process and the number of studies involved [4, 23] as following:

- *Goal*: SMS points to a broader, less concrete research questions than SLR. Its point is to classify items into a number of predefined classes (or clusters). The extracted value is based on the statistical characteristics of the past experience.
- *Process*: The quality of studies is not so important in SMSs, and data extraction methods differ. Additionally, many different methods must be considered to collect more studies.
- *Number of studies*: Consequently, the SMS execution aims to a larger set of primary studies. The classification itself is usually less demanding than the knowledge extraction in case of the SLRs, but it is more important to collect as many as possible suitable studies.

Those differences can be observed in several published SMSs. Some typical examples include two SMSs on DSLs [9, 24], an SMS on aspect-oriented model-driven code generation [25], an SMS on web application testing [26], and an SMS on software product line testing [27].

Because of the important differences between SLRs and SMSs, several authors described their experience and inability to efficiently use the SLR guidelines while conducting an SMS [5, 28, 29]. Consequently, modified guidelines for SMS in SE evolved [3]. The main differences include the additional use of methods for article collection (e.g., snowballing, manual search), adaptive classification and enhanced visualization of results (different types of charts) [23]. Additionally, the guidelines [4] might be too restrictive in certain parts and they limit the amount of suitable literature.

In [3], Peterson summarized the SMS process in four major phases:

1. Need for the map
2. Study identification
3. Extraction and classification
4. Study validity and presentation

All the phases (and contained steps) are shortly presented in Sections 2.3.1–2.3.4. The emphasis is on the differences from the SLR process as described in the guidelines [4], so it is related to the contents of Section 2.2.

2.3.1 Need for the map

Similarly, as in case of SLR, the need for an SMS is identified and the review is planed first. All decisions on conducting SMS are taken at this point. Arksle in [30] stated that the typical research goals are different from SLR and include :

- to examine the extent, range or nature of research activity,
- to determine the value of undertaking a full SR,

- to summarize and disseminate research findings, and
- to identify research gap in existing literature.

This is represented in the form of research questions, which are less specific and oriented toward the knowledge on a specific subject.

2.3.2 Study identification

Next, the studies for the SMS are collected. A number of complementary methods are used here to assure the best possible coverage of available empirical evidence. Then the included studies are selected by screening. The following phases can be distinguished:

1. *Choosing search strategy* should enable a wide enough gathering of evidence. Obviously, the main source is database search. Manual search is beneficial as well as snowballing, so both must be performed. Wohlin advises the following principles for an effective article search [22]:
   - obtain articles from different clusters (nonconnected communities),
   - start with a reasonable (i.e., not too small) number of articles,
   - include different authors, years, and publishers, and
   - define search using keywords from research question.

2. *Developing the search* phase may be performed using different approaches. Using PICO(C) strategy is one of them, but the experience shows it might be too restrictive in case of SMSs. Use of experts proved to save effort later. The search should be improved in an iterative way, and the use of standards and encyclopedias improves the results. Special consideration is needed when dealing with precision (noise) in case of SMS.

3. To *evaluate the search* several usable proposals are available. The aim is to determine the quality of the collection and consequently when the search can be stopped. The paper test-set or test–retest method might be used [3]. The expert opinion is the typical method for this. Petticrew in [21] proposed to stop the database search when no further article can be manually found.

4. *Inclusion and exclusion* phase defines the inclusion and exclusion criteria. In case of SMS, they should be less restrictive to see recent trends. To improve the reliability of inclusion and exclusion process, three strategies have been identified, namely *identify objective criteria for decision, resolve disagreements among multiple researchers* and *define decision rules.*

5. *Quality assessment* should follow one or more of the approaches identified by SR protocol, but in practice, this is a rare case in SMSs due to more general requirements.

2.3.3 Extraction and classification

In this phase, the required data are extracted and classified. The classification is an SMS specific. Several types of classifications are usually present in SMSs.

1. *Extraction and classification process* is usually performed either by using several researchers who check the same evidence or by using a pilot set of evidence to assess the objectivity of criteria. In both cases, the objective criteria for the decision have to be identified. The information that could bias the decision has to be obscured. All disagreements among multiple researchers have to be resolved.

2. *Topic-independent classification* is generally applicable and thus desirable to be used. For an SMS, the most common types of classification are by the venue, by the research type or by the research method.

3. *Topic-specific classification* [23] uses keywords from paper's abstracts to create classification scheme. In this manner, the articles with the similar topic might be found.

2.3.4 Study validity and presentation

Finally, the SMS is validated and the results are presented in suitable forms. The visual presentation is very important for the SMS.

1. An SMS uses several types of charts to improve *visualization*. Most frequently, bubble plots, bar plots, and pie charts are used as they are most suitable to present the numbers related to categorization. Other charts (e.g., line diagram, Venn diagram, heat map) are used as well.

2. The most important *validity threats* for SMS are high publication bias, poorly designed data extraction, high researcher bias, the low quality of sample studies obtained, bad aggregation of the results of mapping, and the low reliability of conclusions drawn.

2.4 Observations on SR

In many years of evidence-based research in different research areas, important lessons were learned. Main findings resulted in advanced approaches described in guidelines and special tools used by many. Still, there are considerations, which are sometimes case dependent and point out the important practical issues.

2.4.1 Lessons learned from conducting SRs

Due to extensive experience in the field of medicine, important experience on conducting SRs is gained in SE as well. Several authors discussed issues and SE specifics arising from the literature. Here, the aim is to point out some of those and present a more critical view of the SR procedure as well. In the following list, a selection of lessons is briefly presented by the SMS phases:

(a) *Need for the map*:
- SR in SE is likely to be qualitative in nature [4].
- We need to search many different electronic sources; no single source finds all of the primary studies [23].
- Current software engineering search engines are not designed to support systematic literature reviews. Unlike medical researchers, SE researchers need to perform resource-dependent searches [6].
- The extent and completeness of the set of empirical studies addressing a particular topic in SE has a small scale and is difficult for classification [28].
- The use of an SMS for student projects at both undergraduate and postgraduate levels is less successful [28].

(b) *Study identification*:
- There are alternative search strategies that enable you to achieve different sort of search completion criteria. You must select and justify a search strategy that is appropriate for your research question [6].
- All systematic review team members need to take an active part in developing the review protocol [4].
- A prereview mapping study may help in scoping research questions [31].
- Expect to revise your questions during protocol development, as your understanding of the problem increases [6].

(c) *Extraction and classification*:
- Piloting the research protocol is essential. It will find mistakes in the data collection and aggregation procedures. It may also indicate that you need to change the methodology intended to address the research questions [6].
- Data extraction is assisted by having data definitions and data extraction guidelines from the protocol recorded in a separate short document [4].
- Review team members must make sure they understand the protocol and the data extraction process [6].
- Having one reader act as data extractor and one act as data checker may be helpful when there are a large number of papers to review [6].

- Review teams need to keep a detailed record of decisions made throughout the review process [6].

(d) *Study validity and presentation*:
- All the medical standards emphasize that it is necessary to assess the quality of primary studies. However, in SE the quality assessment depends on the type of SR you are undertaking [6].
- There must be a separate validation process and a separate protocol piloting activity. Ideally, external reviewers should undertake this validation process [6].
- It is important to be sure how the quality assessment will be used in the subsequent data aggregation and analysis [6].
- The main threat to validity in SR is the presentation bias [28].
- Even when collecting quantitative information it may not be possible to perform a meta-analysis of SE studies because the reporting protocols vary so much from study to study [6].
- Tabulating the data is a useful means of aggregation but it is necessary to explain how the aggregated data actually answers the research questions [6].
- The SE community needs to establish mechanisms for publishing SLRs which may result in papers that are longer than those traditionally accepted by many SE outlets or that have appendices stored in electronic repositories [6].

2.4.2 Practical considerations

In case of the SR process, we may divide practical considerations into the *subjective* (i.e., related to the SR performer) and the *objective* (i.e., related to external factors).

A strict adherence to the SR guidelines and developed protocol ensures to performer the correct definition of the research questions, the performance of the article collection/selection and the data extraction. A lot of experience and effort is needed to follow the prescribed procedure in the expected way, but this part may be mastered by a performer. Still there are some external practical issues, which are much less expert dependent, but they influence the process and the outcome of the SR importantly:

- *The lack of available evidence in digital libraries (DL's)*: the digital libraries contain many, but not everything. Other approaches have to be used, but they have practical limitations as well.

- *The large quantity of available evidence*: more frequently, digital libraries contain a huge amount of very versatile evidence. The required effort may be simply too large to properly review all the evidence.
- *The quality of available evidence*: Important part of SR guidelines is concerned with this issue. Still, a considerable additional effort is required to address this problem.
- *Tool support*: Contemporary SRs are performed mostly using DLs. DL search engines have issues [6]. Similarly, other tools (i.e., reference manager) have problems with consistency, etc.

In all cases, the quality of findings suffers as not the entire corpus of evidence is processed. In the first and in the last case this is due to the missed part of articles and the performer is helpless here. However, in the second and the third case performers are forced to (somehow) limit the amount of the reviewed articles to enable the SR to be performed in available time. Here additional support would be beneficial.

This situation is more frequent in the case of the SMSs than SLRs. As observed before in Section 2.2, the amount and diversity of evidence is larger in the case of SMS as the research questions are broader. Additionally, the selection of included articles is less demanding as they are classified only.

3. Effective article screening: The problem and the solution

The following part presents the core of this article. First, the actual problem is systematically presented. Second, the solution is described in a sequence of subsections addressing topics from basic idea to solution specifics (i.e., decision principles, statistical analysis, and automatic assessment).

3.1 The problem: Manual screening issues

This article addresses the issue of effective article screening, which is an important (maybe even crucial) problem in the practical execution of SR process. Once all potentially suitable articles are collected in a proper way, the subset of actually suitable articles is determined by article screening. In case of a large number of articles, the screening requires a considerable effort. For instance, Syriani reported nearly 70 full working days were needed to perform the screening of 11,173 articles [8]. If the required effort exceeds the available resources, the SR executors tend to limit the number of collected articles or perform screening in a faster way. In both cases, the quality of result (i.e., the set of the included articles) is lower.

3.1.1 The cause: The huge amount of collected articles

Following the SR procedure, the research topic and related research questions implicitly determine the amount of existing evidence by the principle: the wider research topic means the larger set of evidence, which should be collected from different sources. Due to the huge amount of the available evidence provided by the contemporary archives, even for a quite narrow and specific topic, a large amount of possibly suitable evidence is usually collected. This is a result of an adherence to the SR procedure when the collection is carefully conducted and finds a majority of all important articles from all available sources.

A large quantity of evidence is an excellent outcome if all articles are actually appropriate for the research topic. Unfortunately, this is seldom the case in practice. For instance, in case of (most frequent) DL searching for evidence, the research topic and questions are translated into the appropriate search string to be used in the DL searching. Too specific search string excludes important evidence, which is incorrect. If the search string is less specific, a larger set of possibly suitable articles is found. As a consequence, a higher number of inappropriate articles is included as well.

The further SR process is importantly driven by the amount of the collected evidence: a larger set of collected articles demands proportionally more effort than a smaller one. In case of article selection, the effort actually corresponds to more manual work and time.

Additionally, the selection of included articles demands more steps (i.e., phases) in the case of bigger evidence as well. Instead, one browsing through all the collected set, the articles titles and abstracts are quickly screened to exclude obviously inappropriate articles first. Next, the articles are read entirely to exclude some more. Finally, they may be read once more by another SR performer to come to the final decision.

3.1.2 The consequence: The limited SR

Too large set of collected articles is not doable in practice. In one way or another, the resources (e.g., number of the experienced screening performers or time) are limited. Additionally, a large set of articles must be processed by several SR performers, where the problem of same screening criteria and communication arises. Consequently, the SR performers adjust the SR process to be able to perform it in practice.

The first possibility they consider is to intentionally limit the initial collection of the evidence. This may be done in different ways. A safe one would be to stop collecting the articles when the predefined margin of error is reached [7]. More frequently, the SR performers limit the article collection on some sources (e.g., some favorite DL's) only. The other usual way is to narrow the research questions (in case of DL search the search string) to diminish the number of hits to a number, which is realizable in a current situation.

Sometimes the rationale behind this actions may be acceptable in case of SLRs. They have usually a narrower research topic, which can be even adjusted in more iterations to fit the requirements. In case of an SMS, the topic is usually wider and the nature of research is to include all important evidence. Thus, the decision to limit the amount of all collected evidence is riskier as it might cause the initial loss of important part of the evidence. As not all evidence is included, the final outcome of the SMS might be compromised.

The actual problem is in this nonselective initial loss of a part of appropriate articles. In practice, if all appropriate articles were collected, later in the process the majority of them would be excluded anyway, but those excluded articles would be carefully selected by screening.

The second possibility is to process the entire set of the collected articles, but in a quicker, less detailed way. In this case, the SR performers screen the articles in an even quicker mode first (maybe by reading their title only) and exclude all obviously wrong articles. Next step is to regularly manually screen the remaining of the articles.

This is better than limiting the search as all possible articles are checked, but it consumes an additional portion of time and it is still a dangerous practice as the titles might be very misleading.

3.2 The solution: The automated screening

Once the problem is clearly understood, our solution is presented. The idea is to implement an efficient automation of the screening process. In this part, the presentation of the solution is divided into separate descriptions of several important principles.

3.2.1 The idea: Efficient automation

Instead of reducing the size of the set of collected articles or perform screening in a negligent way, the screening might be performed more efficiently by using the power of the computer. If it is performed automatically instead of manually, the size of the set of collected articles is a smaller problem and the initial set of articles can remain as large as needed. Most importantly, the computer can do more in a very short time, so the decision can consider more information, not just an article title. Of course, an automatic procedure should guarantee that the exclusion of obsolete articles is done in enough detailed and controlled way.

Unfortunately, the decision which article to include in the set of appropriate articles is a very complex one. The SR guidelines require a sequence of steps to correctly define and perform this task: a definition of the inclusion/exclusion criteria and the appropriate way to manual screen the articles (i.e., by the reading of the title and the abstract at least). Still, the decision is subjective. It is adapted to the topic and article complexity. It requires a deeper understanding of the article contents, its structure and some other attributes (e.g., the publication title, the author, or similar). This is also the reason why this is not an obvious solution practiced by many.

Still, our proposition is to modify this approach and automate the screening part of SR process. Even though the manual screening has some important advantages those might be challenged by alternative approaches in automatic screening. On the other hand, manual screening has also other issues apart exceeded effort, where the automatic screening would be beneficial. One of those is different bias in case of many screening performers, where the quality of decisions on suitable articles is obviously aggravated or even reduced to some simple principles, which could be efficiently supported by an automatic tool as well.

This is especially true in case of the SMS process. Due to a larger set of collected articles, the manual screening is usually done roughly only by reading the title and abstract (sometimes even the abstract is skipped). If more versatile articles are included in the set, the quality control is less strict (see Section 2.2), and the expected rate of included articles is lower. All these characteristics prove a simpler decision in this case.

The automatic screening can be performed on the entire contents of the article and not only on the title (and the abstract). This advantage should enhance the decision and partly replace the absence of article understanding.

To ensure the adequate procedure, the automatic decision must be configured by experts, who actually perform the manual screening. In a way, they somehow program their decision rules into a computer tool instead of using them manually. Such use of decision rules can be considered as their efficient implementation, which is inevitable in case of a large initial set. Even in case of a smaller initial set, this approach saves a large amount of effort (i.e., time).

3.2.2 Basic principles

Our idea is to gradually develop an automatic solution, which enables the automatic screening of the entire contents of all collected articles. The result is an automatic decision, which articles should be included (and excluded) from the further SR process. The solution implements the following basic principles:

- The solution is primarily aimed to support the SMS research as it (usually) considers a larger amount of evidence and simpler decision rules. Still, it can be used for SLRs as well considering the limitation of an automatic procedure.

- The automatic screening can be used for very different research fields, topics, and questions. It is freely configurable and intended to be used by SR performers, who have enough domain knowledge and (manual) screening experience on the specific research topic.

- The solution defines an approach (i.e., the exact steps) to perform the automatic screening. The approach is defined in such a manner it can replace manual screening part in the SR process (performed following the SR guidelines). It is adaptable to different levels of manual involvement and can be used in an entirely automatic, manual, or combined way.

- Additionally, the solution provides a tool to efficiently support the approach. The tool is designed to be as simple as possible to use, and adaptable to other tools the SR performer may use. The input and output are actually sets of files in the PDF format.

- An important solution rule is to be careful, thus not to exclude any possibly suitable article, while the opposite (i.e., to include the wrong article) is admissible. This is not a big problem as the later steps will exclude a wrong article. To support this rule, the articles are classified into three sets instead of two. Apart from the sets of included and excluded articles, a set of possibly included articles is added, too. Here all the margin articles (i.e., those were the decision on inclusion cannot be taken without a detailed reading of the article) are inserted. The actual level of this rule implementation is configurable and dependent on SR performers.

- The efficient screening automation demands a careful configuration of the decision part. To ease this an iterative approach is advised. Initially, an approximate solution is created and this is refined and upgraded in an iterative way later. All the necessary decision-making information is saved and may be reused for later adjustments and upgrades.

- The solution becomes usable even if the configuration is just initialized. At first, all articles might be classified as possibly included. This is a valid, yet not really useful decision. As the construction of solution advances in an iterative way, the set of possibly included becomes smaller. The iterations must be stopped before the level of wrong decisions starts to increase.
- To efficiently implement automatic screening, the idea of adaptive tuning is implemented. To this aim, a pilot subset of an entire initial set of articles is used. Here the results of the automatic screening can be assessed in comparison with the correct (i.e., manual) results. In this manner, the automatic procedure can be more efficiently improved in an iterative manner.
- All the data the approach uses and provides are saved in a predefined way using standard file formats. For instance, the articles are saved in the PDF format, the configuration in XML format and the report in the XLS format.

Considering all basic principles, the outline of the automatic screening solution is defined. Its basic elements are the following:

(a) *INPUT*:

Two types of data are needed to conduct the screening: an initial list of all collected PDF articles, and a configuration, which includes all the parameters to define the file locations, pilot screening and criteria for the decision-making.

(b) *ACTION*:

The automatic screening reads the parameters from the configuration, prepares the environment, and performs the screening in two possible ways: as a pilot or as main (i.e., entire corpus) screening. Additionally, it should enable the insight into screening outcomes and their manual manipulation.

(c) *OUTPUT*:

The outcome is a report with screening results and three lists of articles: the included, the possibly included, and the excluded articles.

3.2.3 The decision approach

Before continuation, we have to decide on how to implement the decision-making mechanism, thus which type of the article analysis is to be performed and how the decision rules on included and excluded articles are set. In our view, two different approaches may be used to implement decision for the automatic screening.

A simpler one is to implement a form of text mining. In our case, we consider the text statistic analysis, where the frequency of selected words (i.e., keywords) is determined. Then, the decision important article characteristics are recognized from the obtained numbers directly. In this way, the decision principle simulates the manual screening situation, when frequent uses of specific words demonstrate a important research topic is actually addressed in the article. Thus, the domain expert is needed to determine the decision-making setting. For instance, different visual text mining solutions were proposed [11, 12].

Another solution is the implementation of artificial intelligence approaches (for instance automatic learning using naive Bayes, ontology-based approach, etc.). This solution uses the same text statistic analysis data, but it uses the principle of knowledge extraction to define the decision rules. A learning set of previously defined positive and negative manual decisions is used to teach the system how to decide, thus the decision rules are not directly set by the human decider. Solutions using the advanced AI principles were proposed in the research area of medicine [32–34].

At first, the first approach seems inferior as it relies on the domain expert's capability to define the best setting, while the second implements a larger portion of actual automatic decision-making. A rule of thumb is that the implementations of the simple approach work quite well, but they might not cope with the complexity of the task and yield worse results sometimes. On the other hand, the complex AI solutions need a lot of data for efficient learning and tend to be too complicated to be efficiently used in practice.

There are other considerations, too. An important view of the solution approach is how understandable is the decision-making mechanism. Using the first approach, the decision adjustment is simpler but very dependable to the specific research topic. In some cases, the decision rules can be simple and straightforward, while in other cases they might prove to be complex. But even if the first approach tends to require more effort from the domain experts, it produces a more understandable set of decision rules.

The AI solutions require less setting, but the final decision rules can become very versatile and without a logical explanation as they implement the principle of automatic learning from data. The quality of the solution is very dependable on the quality of the learning set, which may be a problem. The experience proves, AI solutions still base on a topic-specific understanding of the decision process, which is implemented in their core. As an example, the solution [33] rely on specifics from evidence-based medicine and it is not directly usable in other environments (e.g., SE).

Both approaches base on the text statistics and they may be seen as two alternative decision approaches, where the first defines straight automation of manual decision-making, while the second involves a computer to derive decisions from data. In the first case, the quality of final outcome (decisions) relies more on human involvement, where the computer tool empower its efficient implementation. The solution can be gradually upgraded through an evolving definition of decision rules by the human decider. This approach is simple and it fits with our concept of automatic screening.

An increasing quality of the learning set is needed to upgrade the quality of solution in the second case. Consequently, the learning set (i.e., the set of decisions of manual screening) has to grow bigger and all the decisions are considered correct. The AI approach is a riskier solution for the automatic screening as the decisions are (somehow) steered indirectly and the outcome is less certain. This yields a more versatile process for performers, which diverges from our screening principles. Still, we see this approach as an alternative for possible future implementations.

Both approaches might be supported in an automatic screening process, but in our case, only the first is currently fully covered. The following requirements are needed to implement the decision solution:

- First, a detailed text statistic analysis has to be carefully implemented.
- Next, the iterative nature of our process already supports a gradual improvement. The efficient implementation of decision rules is needed, too.
- Finally, our concept of pilot screenings address well the requirements of evolving learning set.

3.2.4 The text statistic analysis

Following this rationale, the text analysis has to be performed to determine the statistical characteristics of a set of selected *keywords*, which may be members of different *groups*.

The specific keywords have to be selected by the domain expert, possibly the actual manual screening performer. The screening performers have a lot of experience in choosing the right keywords as they browse through the text (actually the title and abstract) and search for certain characteristic descriptions connected to a research topic. Those are usually tightly related to a set of selected keywords.

The entry step of text analysis is to define such a set of keywords and count the number of occurrences of each keyword in the article. The counting may be performed using a strict or a loose match of keywords.

For example, word "processes" is not a strict match for the keyword "process," but it is a loose one.

Additionally, each keyword is a member of one group. In the extreme situation, all keywords constitute one (i.e., same) group, but they can be also placed each to its own group. The number of occurrences is counted for the group in the same manner as for a separate keyword—cumulative of all occurrences of group keywords defines the group count.

There are other (more complex) characteristics which could be counted: the number of keyword occurrences in a specific part of the article (e.g., title or abstract), the relation between different keywords, etc. Those characteristics demand strict adherence to the article structure and encumber the decision-making. As the generality and the ease of use is required for our automatic screening, the text analysis is limited to the basic counting of keyword and group occurrences.

3.2.5 The decision-making

The collected statistical results are the base for the implementation of the direct decision-making. Some observations on the relation between the selected keywords occurrences and the article inclusion decision are discussed before the description of the decision-making mechanism.

- *The required keywords* Usually, the screening performer defines keywords, which have to be present (one or many times) in the text. The absence of such keywords is a clear indication the article is not suitable for a specific SR. An example of such a keyword is "review" while conducting a tertiary SR, which analyses a number of different reviews.

- *The negative keywords* Manual screening quickly excludes the article when the performer discover a description of the wrong topic or even wrong research area, which is characterized by the presence of certain (forbidden) keywords. This is a frequent case as the acronyms are used as part of the selection string during the process of article collection from DLs. As an example, the same acronyms may have a different meaning in different research areas (e.g., the acronym DSL means domain-specific language in computer science, but digital subscriber line in Telecommunications).

- *The positive keywords* Sometimes it is possible to define a keyword, which is not required, but when it is present, it clearly indicates the article is suitable for a specific SR. For instance, the article with a keyword "systematic literature review on DSL" is definitely an included one while conducting a tertiary SR on DSLs.

- *The groups of keywords* In most cases, the situation is not so obvious, but there is a number of keywords, which indicate the article might be included. The presence of one is not enough, but the occurrence of more keywords indicates the article describes a specific research topic. When such keywords constitute the same group, the decision can be based on this group's statistics.

- *The number of occurrences* In case of topic-specific keywords, even a small number of occurrences prove article connections with the research topic. In case of commonly used words, this is not the case, but a very frequent use of specific keywords is still indicative. Thus, the number of occurrences is important for inclusion decision, too. For instance, even one use of a keyword "domain-specific language" demonstrates the article is connected with that topic, while there should be a lot of occurrences of a keyword "domain" or "language" to determine the same.

- *The synonym keywords* In practice is difficult to find only one keyword which is consistently used in all articles. Instead, a group of keywords may represent the same meaning. To include all relevant articles, the use of any alternative must count. The implementation of this principle is possible with the use of groups. For instance, a group might consist of keywords "domain-specific language," "DSL," and "small language."

- *The alternative groups* When the articles cover distinct but connected topics, the principle of alternative groups of keywords might be needed. The article is included in the first or second group indicate so. An example of this principle is the research topic on DSL and DSML. The first group consists of keywords "domain-specific language" and "DSL," while the second includes keywords "domain-specific modeling language" and "DSML."

Putting it all together, a decision-making mechanism can be designed in a simple, yet flexible way. The decision rules are set for the keywords and groups. For each keyword, a required, negative and positive number of occurrences can be set (but it is not required). The same is defined for each group of keywords. Additionally, a logical expression combining the decisions based on the statistic characteristics of separate groups defines the final exclusion criteria.

In the beginning, an article is considered possibly included and the number of occurrences of each keyword is sequentially checked. Each keyword can be decisive for the exclusion or inclusion of the article. Table 1 presents the decision criteria, which are based on the preset required, negative and positive number of occurrences for all the keywords (and groups).

Table 1 The exclusion and inclusion criteria decisions.

Criteria	Defined on	Number of occurrences	Decision
Required occurrences	Keyword, group	Less than	Exclude
Negative occurrences	Keyword, group	Equal or more	Exclude
Positive occurrences	Keyword, group	Equal or more	Include

If the decision is not taken after processing all the keywords, the group based decision applies (sequentially for all groups). Thus, the number of cumulative occurrences of all group keywords is used to decide on the exclusion and inclusion in the same way.

If also this step fails to give a verdict, the final decision on article exclusion can be a result of a logical expression between groups decisions. In some cases, all decision steps do not yield a definitive answer.

The direct decision-making can be efficiently performed once the statistical results on the keywords and groups are gathered and the decision rules are (appropriately) set. The decision places the article in one of the three sets: included, possibly included and excluded. The characteristics of the sets are the following:

1. *Included articles*: they have a required number of occurrences for all the keywords and groups, less than negative number for all negative numbers of occurrences for all the keywords and groups, and at least one situation, when they have equal or higher number of occurrences than demanded by a positive number of occurrences for a keyword or group.
2. *Excluded articles*: they do not reach the required number of occurrences for at least one keyword or group, or they have equal or higher number of occurrences than defined by at least one negative number of occurrences for a keyword or group.
3. *Probably included articles*: the remaining articles.

Example. Consider the screening is using the decision criteria presented in Table 2. The first two keywords from the group "In" are expected to be present in the article, while the presence of keywords from group "Out" indicates the article might not be suitable for the specific research topic. Thus, the logical expression for exclusion is "In AND NOT Out."

In Table 3 the actual results (i.e., the number of occurrences of keywords) for three sample articles are presented. The Article 1 is excluded as it does not contain the required number of occurrences of keyword "DSL" (2). The article 2 contains enough keywords "DSL," but it is also excluded as it contains too many occurrences (7) of negative keyword

Table 2 An example of decision criteria with groups "In" and "Out" each containing two keywords.

Keyword	Group	Required	Negative	Positive
DSL	In	2		5
Language	In			5
Digital	Out		2	
Ontology	Out		2	
	In	2		5
	Out		2	

Table 3 The actual number of occurrences for three sample articles.

Article	DSL	Language	Digital	Ontology
Article 1	1	0	1	0
Article 2	2	0	7	0
Article 3	4	3	0	0

"digital." The Article 3 is included as it does not violate any exclusion criteria (considering the logical expression, it has enough required occurrences of "DSL" without any occurrence of keywords from a negative group "Out") and it contains more than defined a positive number for keywords from the group "In."

A variety of different automatic screenings can be performed using this decision mechanism. When a small number of selected keywords is used, grouped in a single group and the decision rules are loosely defined, the screening might repeat the DL search (with the same search string). In this case, all articles are possibly included and the screening is trivial.

A strict definition of decision rules (i.e., required and negative numbers of occurrences) on the same keywords might yield a decision when the majority of articles is excluded or included, but in a very nonselective way.

By adding more keywords and groups the mechanism becomes more sensitive but more difficult to use as well. A balanced use of keywords and groups with the logical decision rules should give the results, which are comparable with those of manual screening.

In theory, using a large number of keywords and groups with complicated decision rules it is possible to achieve a result, where all articles are correctly divided between included and excluded sets. Unfortunately, this is quite difficult to achieve in practice.

The decision-making must be carefully adjusted to specific research topic up to a reasonable point when the expected results are achieved with a suitable effort involved. Obviously the efficient screening configuration bases on clear inclusion/exclusion criteria. It requires an experienced performer with deep understanding and insight into the research topic. Even though, the appropriate configuration is a result of several iterations of adjustments. Each configuration has to be tested on a small (pilot) portion of articles using a comparison with the manual screening decisions.

3.2.6 The automatic decision assessment

Two factors importantly influence the practical efficiency of our automatic screening: (a) the portion of manual screening required to perform pilot screening and (b) the effort needed to define efficient decision rules.

(a) Obviously, the idea is to manually screen as little as possible articles and to still define the decision rules correctly. The practice proves that this is rarely the case due to versatility of collected articles and manual screening issues (e.g., decision using article's title and abstract only, decision bias of different screening performers). Unfortunately, our solution can assist here with an effective pilot screening management only.

(b) On the other hand, the definition of decision rules is an iterative process. The goal is to achieve a reasonable definition of rules in a limited number of iterations. To gradually improve the successive decision rules, the current version has to be assessed. An automatic assessment and a clear presentation of the current decisions on included/excluded articles is a valuable support to the screening performer.

To enable the automatic assessment, the referenced (i.e., manual) decisions have to be prepared in advance. This requires a slightly different organization of pilot screening as the pilot articles have to be selected from a set of already manually screened articles. Additionally, the results of such a screening have to be presented in a suitable manner for further assessment.

After the pilot screening, the decisions of the automatic screening are compared with the referenced decisions of manual screening and marked with a numeric mark. The marking expresses the conformance of the automatic results with the referenced ones.

The marking is based on three principles. First, in the optimal case, all manual and automatic decisions should be the same. Second, if they are not, it is better to decide (i.e., include or exclude) than not to. Finally, the decision to include an article is better (safer) than the decision to exclude.

Table 4 The marking of the referential and automatic decisions.

Automatic screening	Referential screening		
	Include	Possibly include	Exclude
Include	0	1	6
Possibly include	4	3	5
Exclude	7	2	0

The referenced (manual) screening of a specific article results in three possible outcomes (i.e., to include, to possibly include or to exclude), and the same is true for the automatic screening. Table 4 presents the nine different comparison possibilities, which are marked according to the following rules:

- *The two "perfect matching" alternatives*: the decision to include/exclude is taken, and the referenced and manual decisions are the same (mark 0)
- *The two "decision taken" alternatives*: the manual screening cannot decide and yields possibly include, but the automatic assessment decides to include (mark 1) or exclude (mark 2)
- *The "no decision match" alternative*: both the referential and automatic screening cannot decide (mark 3)
- *The two "no decision" alternatives*: the automatic screening cannot decide, but the referential decision is to include (mark 4) or exclude (mark 5)
- *The two "mismatch" alternatives*: both decisions are taken, but they totally mismatch—the automatic decision is to include (mark 6) or to exclude (mark 7)

The marking is clearly presented and enables the screening performer to quickly locate the articles with the worst mismatch. The marks may be used to gain an overall (i.e., average) assessment of the decision quality.

Additionally, the automatic screening should yield a decision (i.e., to include or exclude) for most of the articles. A delicate automatic assessment might yield possibly included for most of the articles, but such a result has no practical value. By calculating the percentage of decided screenings enables the missing insight into the quality of automatic assessment.

Example. In Table 5 the decisions of the manual and automatic screening for five sample articles are presented. The marks obtained are located in the last column. The final average mark (2,8) proves the automatic screening decisions differentiate significantly from the manual ones. The decisions to include or exclude are taken in 40 % only, thus the automatic screening has to be improved further.

Table 5 The marking results for five sample articles.

Article	Referential screening	Automatic screening	Mark
Article 1	Include	Include	0
Article 2	Possibly include	Exclude	2
Article 3	Include	Possibly include	4
Article 4	Exclude	Possibly include	5
Article 5	Possibly include	Possibly include	3
Average			2,8

3.2.7 The iterative adaptation of decision rules

Once the decisions are obtained and marked, the gained knowledge is used to upgrade the configuration. The principle is to inspect the screening report, detect the wrong decisions and try to define a decision rule, which would classify the article in the right set.

The principles presented in Section 3.2.6 use the required, positive or negative number of keywords (and groups) found for the definition of exclusion or inclusion criteria. Consequently, each keyword or group may be checked in the report to find out, if each number of occurrences larger than a certain threshold value actually coexist with the same (i.e., included or excluded) decision of manual screening. In this case, we can assume, which by setting an exclusion or inclusion criteria on a corresponding threshold value, a better final outcome might be found.

This principle should work perfectly when inclusion/exclusion criteria are strictly aligned with the use of specific keywords. In this case, it is possible to obtain a clear distinction on included and excluded part on the list of articles sorted by one of such keywords.

Unfortunately, this is not frequent as the decision is based on a deeper understanding of article contents. Additionally, the reviewer's bias may also cause wrong decisions. Considering this, the setting principle of decision rules might be loosened up and ignore the articles, where a manual decision was not taken (i.e., classified as possibly included).

Furthermore, such setting may skip one (or some) articles, which are oppositely classified as well, when the majority of articles at the beginning (or at the end) of a sorted list has the same decision. By setting such an imprecise rule we consciously classify some wrongly to achieve a larger part of articles, where the decision is actually taken. Table 6 presents possible ways for upgrading decision rules.

Table 6 Alternatives for upgrading decision rules.

Adaptation	Description
None	The decision rules are fixed
Delicate	All articles with more than X occurrences have the same decision
Reasonable	Same as before, but possibly included decisions are ignored
Strong	Same, also up to three articles with opposite decision in row are ignored
Radical	Same, but up to five articles with opposite decision in row are ignored
Brutal	Even more than five articles with opposite decision in row are ignored

When five or even more articles with opposite decision are ignored, the decision rules become very loose and a significant number of articles is classified incorrectly. In this situation, most or of the articles are classified predominantly in the exclusion set.

This should not be the case when using consistent manual screening decisions based on clear inclusion/exclusion criteria. But as we perform upgrading using the information from a (small) pilot set, the rules may not yield good results. Using a larger pilot set is beneficial here, but this demands more work for manual screening.

Example. When the articles are sorted by the number of occurrences of a specific keyword "DSL" and all articles with less than a specific number (5) are manually excluded, a clear indication for a new (or changed) decision rule is present: the required number of occurrences for this keyword has to be set at 5. In this way, a "delicate" adaptation is applied. By applying the "reasonable" adaptation, the required number of occurrences of keyword "DSL" might be set at 10, if all the articles with 10 or fewer occurrences yield the manual decision excluded or possibly included. With other adaptations, this number might be set to even higher value.

4. The implementation of the automatic screening

Two artifacts are developed to implement the before mentioned solution. First, a rigorous process to efficiently perform the automatic screening is defined. Second, a specific tool to support the process execution is developed as well.

4.1 The automated screening approach

The proposed solution is implemented and integrated into a modified SMS process, which now relies on the automatic screening. The approach strictly follows the SMS guidelines [3] till the collection of all studies from the different sources is finished. At that point, the manual screening part of the process is replaced with the automatic screening process, which is conducted in the following way:

1. *The configuration*

    The tool for automatic screening has to be configured by an expert (usually the intended performer of the manual screening) before the process can be performed further.

    First, the locations (i.e., the directories) of articles are defined for the initial, the referential and the final set of articles. The location for the initial set of articles actually contains two separate subdirectories: one for articles in the PDF format and the other for the articles in the TXT format. Next, the parameters for the pilot processing and reporting are set. Then, the assessment part of the tool is configured by the definition of a selection of the appropriate keywords and groups. Finally, the decision rules are set by the depiction of two value for each keyword and group: the exclusion criteria defines the required and a negative number of keyword/group occurrences in the article (one positive or negative value suffices as both values cannot coexist) and inclusion criteria defines a positive number of occurrences. The configuration is saved in a file in a self-descriptive XML format.

2. *The preparation*

    A few preparation steps are needed before the actual screening begins.

    First, a selected configuration file is used to initialize the tool. As the predefined article locations (i.e., directories) may not exist, the tool creates those directories. Next, the initial (i.e., previously collected in the SR process) and the referential (i.e., manually screened) articles in the PDF format are placed in the predefined locations. As an assistance, a reference management tool may be used (e.g., Mendeley,[b] OneNote,[c] JabRef[d]) to manage a complete list of the articles details. Such a list

---

[b] http://www.mendeley.com.

[c] http://www.onenote.com.

[d] http://www.jabref.org.

may be already prepared if the reference management tool was used in the previous collection phase of the SR process. Finally, all articles in the initial set have to be translated from the PDF format into the TXT format and placed in the predefined location for articles in the TXT format. Our tool performs this operation automatically, but an external translator tool (a PDFtoTXT translator[e]) can be used to manually manage the set of articles in TXT format, too.

3. *The pilot screening*

The tool usually performs at least one pilot screening to tune up the screening process before the complete article screening is run.

First, the actual size (i.e., the number of articles) of the pilot set is read from the configuration file. Then, a separate pilot environment has to be established (for each pilot execution). It contains the locations for initial and final sets of the pilot articles in the TXT format. Next, the pilot articles are randomly selected from the initial set of articles and copied into the initial pilot location. Finally, the pilot screening is executed on the articles in the TXT format. The results are three separate sets of the piloted articles (the included, the possibly included and the excluded) located in three separate locations. These lists of articles may be presented and/or saved in several ways. The most useful form for the further screening tuning is a spreadsheet file containing the detailed statistical results for each keyword and group. Also, PDF files are accordingly moved into three separate locations: the included, the possibly included and the excluded articles.

4. *The pilot assessment*

The expert assesses the results of the pilot screening to establish the quality of automatic screening.

This part can be conducted automatically or manually. In the first case, the decision part of the tool uses the existent referential decision for each article and assesses the quality of decision made by the tool (see Section 3.2.7). In the second case, each article from the pilot set is quickly manually screened. The manual decision is compared with its automatic screening result to detect issues. The manual screening is thus performed during the pilot assessment step and not previously as in the first case. The manual approach does not provide the overall mark, but it enforces better understanding of the reasons behind tool's wrong decisions.

---

[e] http://pdftotxt.com.

5. *The adjustment*

If the overall pilot results are unsatisfactory, the tool has to be reconfigured (i.e., the decision part of the configuration changed) and the pilot screening repeated.

The tool enables the repetition on the same or on the different set of the pilot articles, which might have the same or a different size. The adjustment of the configuration can be repeated many times usually resulting in the increasingly higher quality of the outcome each time. Each adjustment is saved as a next pilot screening on the separate locations, containing all the information on the assessment and decisions.

6. *The main screening*

Once the expected quality of the automatic screening is achieved, the automatic screening is performed on the initial (i.e., entire) set of collected articles in the TXT format.

At this point, the main screening environment is set and the screening is performed using the best decision configuration from all the pilot screenings. All the results are saved in the predefined ways, too. The most important results are separated sets of the included, the possibly included and the excluded articles in the TXT format. Additionally, the main screening can be repeated several times using different configurations and result's locations as well.

7. *The verification*

The decisions have to be verified and validated by experts to establish the actual quality of the automatic decision.

When the automatic assessment is used during the decision configuration improvement, the results may be considered verified as the procedure was performed consistently on entire pilot set. In case of manual assessment, the results should be verified as the level of the manual performance may vary. In both cases, some form of validation is needed. The experience proves the aggregated values of the outcome (i.e., the sizes of final sets) and the manual check of some randomly selected articles is beneficial. In case of any changes, the manual manipulation of the results (i.e., moving the articles from one set to another) is allowed to fine tune the final result.

8. *The conclusion*

The inclusion/exclusion decisions are now final and the results are available for the further use.

All sets (i.e., the included, the possibly included and the excluded) of articles in the TXT format are accordingly copied in the PDF format to the predefined separate locations. The tool provides appropriate exports so the resource management tools can import and use the new information. The final step is to archive all the screening results in an economical or in a full (i.e., complete) way. The economical archive saves the information on the configurations and results only. Thus, all the locations with a large number of (PDF or TXT) files are omitted as they contain different distributions of the same articles. The full archive creates a ZIP file containing all the screening locations and files.

At this point, the SR process returns on the usual track. The SR process is strictly driven by the appropriate SR guidelines.

Our screening approach is presented in Fig. 1. Apart from the unavoidable process configuration performed by the screening expert, it includes an actual manual intervention (i.e., article screening) at two points: at the pilot assessment and at the final verification step. More manual involvement is beneficial, but not required. Thus, the level of the manual effort may vary from nothing to everything. In the first case, the pilot screening is omitted (i.e., empty pilot set) or the automatic assessment is performed using the before provided referential data on screening from elsewhere. In this case, the validation only checks the aggregated data sometimes. The other extreme is the screening when all the collected articles are put in the pilot set, thus all articles are manually screened. Such a configuration is disputable as it somehow duplicates work if we consider the manual screening correct. Yet it can be used during the screening process/tool improvement. Usually, the moderate level of manual intervention is required.

4.2 The automatic screening tool

A special tool was developed to provide assistance for the screening performer during the automatic screening process. It incorporates all of the solution principles described in Section 3.2 and strictly follows the automatic screening process.

The tool has three modules (see Fig. 2) to perform three separate sets of activities: (a) the *CONFIGURATOR* is used to set the article locations, screening parameters and decision rules, (b) the *PERFORMER* prepares the environment, transforms the articles in the required TXT format, actually performs the screening and saves the results, and (c) the *MARKER* assesses the screening results and presents the marks. The modular tool design enables the addition of other activities later.

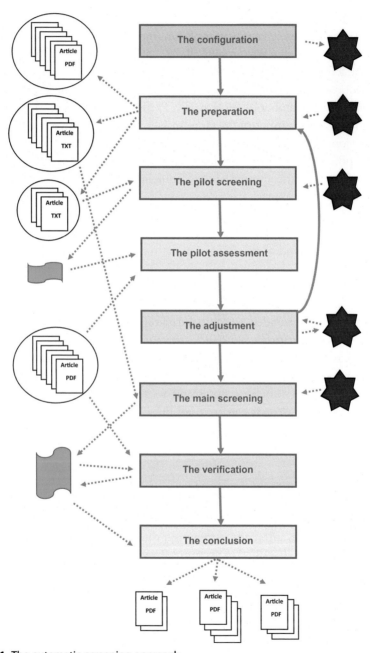

Fig. 1 The automatic screening approach.

Fig. 2 The modules of the automatic screening tool.

The tool is implemented as a configurable Java application. To be generally usable, it supports the principle of standard format's use for all the inputs and outputs. This enables the cooperation with any tool the screening performer may use.

(a) *The CONFIGURATOR:*

This module is used to efficiently configure all the necessary settings before the automatic screening. It follows the principle of the default settings: those are used in case of any missing definition. The configuration is saved in an XML file (see a sample[f] in Fig. 3), which is required by the PERFORMER. More precisely, the CONFIGURATOR offers to screening performer the following possibilities:

- *Describes the configuration*:
  - in the configuration file in the XML format.
- *Defines the input and the output locations*:
  - the working directory;
  - the relative locations for an initial set of the PDF and the TXT articles;
  - the relative output locations for three groups of PDF files: the included, the possibly included, and the excluded;
  - the relative location for the screening results and the name of the file with complete screening report.[g]
- *Defines the reference location*:
  - the relative location of reference results, where three groups of PDF files (the included, the possibly included and the excluded) may be found.
- *Defines the pilot screenings*:
  - the number of articles for the pilot screening;
  - the relative location for the pilot directories.[h]

---

[f] The configuration in this specific XML file defines a pilot screening of 1000 articles using two groups (i.e., DL and NEG) and three keywords (i.e., DSL, DSML, and ONTOLOGY).

[g] A number is added to the report's name to separate reports for the same screening.

[h] A number is added to the directory name to separate the locations of different pilot screenings. Each pilot location replicates the relative input and output structure.

```
<?xml version="1.0" encoding="UTF-8"?>
<analysis>
    <description>Test Big A Approx Strong Decision</description>
    <dir>C:\\Users\\Admin\\SMSTool\\BIG\\</dir>
    <pdf>PDF\\</pdf>
    <txt>TEXT\\</txt>
    <log>LOG\\</log>
    <included>INCLUDED\\</included>
    <maybe>MAYBE\\</maybe>
    <excluded>EXCLUDED\\</excluded>
    <reference>MANUAL\\</reference>
    <pilot>1000</pilot>
    <condition> DL AND NOT NEG </condition>
  - <group>
        <title>DL</title>
        <exccg>0</exccg>
        <inccg>50</inccg>
      - <keyword>
            <title>DSL</title>
            <excc>1</excc>
            <incc>40</incc>
        </keyword>
      - <keyword>
            <title>DSML</title>
            <excc>0</excc>
            <incc>30</incc>
        </keyword>
    </group>
  - <group>
        <title>NEG</title>
        <exccg>-1</exccg>
        <inccg>0</inccg>
      - <keyword>
            <title>ONTOLOGY</title>
            <excc>-1</excc>
            <incc>0</incc>
        </keyword>
    </group>
</analysis>
```

Fig. 3 The XML file with configuration.

- *Defines groups and keywords*:
  - a set of groups;
  - for each group a set of belonging keywords.
- *Defines decision rules*:
  - for each group, the exclusion and inclusion criteria representing the required, negative and a positive number of occurrences[i] ;
  - for each keyword, the exclusion and inclusion criteria representing the required, negative and positive number of occurrences;
  - the final decision rule using the logical expression of group's decisions.

---

[i] The required and negative number of occurrences cannot coexist, so they are saved as one value (i.e., exclusion criteria) only.

(b) *The PERFORMER:* Prior to the actual screening performance, the initial set of collected PDF articles has to be copied to a predefined location. Similarly, the referenced screening results are copied to the appropriate locations (i.e., the included, the possibly included and the excluded). Next, the module performs the following actions:

- *Imports the predefined configuration:*
  - uses predefined configuration XML file.
- *Performs the initial file manipulation:*
  - if necessary, creates all configured input and output locations;
  - if necessary, transforms all PDF files into TXT files and copies them to the specific location.
- *Performs the pilot screenings:*
  - prepares the locations for the pilot screening and copies the entire input and output structure;
  - randomly selects the specific number of PDF and TXT files and copies them to the pilot directories accordingly;
  - performs the automatic screening on the pilot set using the defined decision rules;
  - presents the detailed and cumulative statistical results and saves them into a detailed pilot report;
  - copies the pilot PDF articles into three pilot output locations: the included, possibly included and excluded;
  - enables the repetition of the same screening or use of screening results as a reference for the next pilot screening.
- *Performs the main screening:*
  - performs the automatic screening on the entire set using a defined decision rules[j] ;
  - presents the detailed and cumulative statistical results and saves them into a detailed report;
  - copies the PDF articles into three output locations: the included, possibly included, and excluded;
- *Enables the manual manipulation of PDF articles:*
  - may be assisted by a reference manipulation tool.

(c) *The MARKER:* After the automatic screening, this module enables assessment and presentation of the differences between the referential and the automatic decisions. The module performs the following actions:

---

[j] In case of repeated screening, all the results are overwritten except the screening report.

- *Performs the marking of the screening decisions*:
  - the decisions of referential and automatic assessment are compared and marked;
  - the marks are presented and added to screening report;
  - The cumulative marks (i.e., the size and the percentage of the included, the possibly included and the excluded sets, the average mark) are calculated and presented.
- *The results are used to upgrade of decision rules*:
  - the marker eases the improvement of decision rules by presenting the decision data in the sorted order.

5. The experiment description

An experiment is presented to evaluate the proposed solution and demonstrate its usability on a practical example. The experiment is conducted on an actual SMS we are currently performing. The aim of this SMS is to review the available evidence on examining different processes while developing a DSL or DSML. Actually, the idea for automatic screening is based on some actual issues and experience from this SMS.

At the beginning of the section, we provide a short insight into the contents of our SMS. This is necessary to understand the relation between the concrete SMS research topics (research questions, inclusion/exclusion criteria) and the automatic screening configuration.

In the main part, the experiment details are presented. As the phase of manual screening is already finished in our SMS, the actual decisions are used to compare with the ones obtained from automatic screening. Several alternative solutions for the automatic screening are performed and the results compared focusing on four selected issues. The aim is to define an adequate size of the pilot set and demonstrate our principles for the decision rule definition enable suitable results. Thus, our solution for automatic screening can be used as a supplement for manual screening when configured correctly. The experiment results are presented graphically and accompanied by short discussions, which enable insight into the practical use and limitations of the approach.

5.1 Context: An SMS on examining different processes while developing a DSL or DSML

Our research work is focused on the field of Model Driven Engineering (MDE), more specifically on DSLs and DSML. Currently, we investigate the use of formal development approaches in (the industrial) DS(M)L development. The first step is to conduct an SMS and discover the issues and gaps in this field.

5.1.1 MDE, DSL, and DSML

Model-driven engineering (MDE) promotes the development of models at different levels of abstraction. The higher level models are (automatically) transformed into lower level models and finally to the code [35]. The great potential of this approach is to enable the efficient software development at a higher level, where the amount of routine (programming) work is considerably reduced (as in case of [36]).

Domain-specific language (DSL) is a small programming language focused on a specific domain only. As such it is an opposite to general–purpose language (GPL), which tends to be suitable for all purposes [37]. A DSL is designed for a specific domain, thus it is expected to have enhanced usability among domain users, who are usually not programmers. A DSL is developed by a language expert in cooperation with a domain expert. The definition/implementation of the appropriate DSL ease the solution's development and reduces the possibility of errors.

In our case, it is important to distinguish between *grammar-ware* and *model-ware* DSLs. The first ones are designed as an ordinary programming language (i.e., performing syntax and semantic analysis) and described by a grammar [38].

The second ones are based on the model definition. They are represented using meta–model description, which is defined using UML or some existing tools [39, 40]. Such DSLs are also called *domain-specific modeling languages (DSML)*.

By the MDE principles, a DSML and its description (by an appropriate model) can be understood as the two models at the different levels of abstraction. Due to the obvious similarities between DSL and DSML, a DSL and its grammar description can be considered a model as well.

5.1.2 The SMS research topics

Due to increasing interest in the DSL research [41], a variety of different aspects of DSL development have been addressed in a large set of primary studies: from basic construction issues [42–44] to quality concerns [45, 46]. Several SMSs on DSLs have already been performed as well. To name a few, Kosar [9] and Nascimiento [24] published SMS on the actual state of the DSL research, Syriani [8] provided an SMS on template-based code generation (TBCG), while Mendez Acuna [47] performed an SLR on software product lines (SPL) engineering in the development of external DSLs.

Our SMS is designed to provide insight into two important aspects of DS(M)L development: the actual use of the formal process and the role of the user in this process. Additionally, two related aspects are on our agenda: the role of the development approach and the use of the accompanying tools.

- *The process for the DS(M)L development*

  In the DSL/DSML development, the use of appropriate defined process ensures that the expectations of end users from the specific domain will be actually met. There were several attempts to define a DSL process. The decision, analysis, design, and implementation are considered the obligatory phases of DSL development [48]. A specially tailored DSL development process is described also in [49], while [50] presents an incremental version of DSL development process.

- *The role of the development approach*

  Diverse development approaches can be used for the DSL/DSML development: cascade, iterative, incremental, prototype, or agile approach [51]. In view of our research, the prototype and especially agile development [52, 53] one is beneficial as implies the user involvement [45, 54]. Even as the actual level of actual user involvement varies a lot, recognizing implementation of agility is important.

- *The role of the end user*

  End-user decides if a DS(M)L development produced a usable language in practice. Thus it is important that it is actively involved in development [45, 54]. Due to his lack of programming knowledge, this is rarely the case. Research proves that development without user involvement frequently leads to the bad design of a D(S)ML.

- *The accompanying tools*

  The development of accompanying tools gives an interesting insight into the core of a DS(M)L development. The idea is that a usable DS(M)L must be supported by a tool [48]. The concept presentation without an actual implementation usually demonstrates a solution, which was not tested in industry.

5.1.3 The SMS protocol

The further in-depth presentation of SMS's specifics is out of scope for this work. A defined SMS protocol concisely presents the main characteristics of our research. It was designed by the principles provided by Kitchenham [4].

Only a short outline of most important parts provides enough insight into our SMS. The main outlines from the protocol are:

- *The research method*

  We carefully follow the SR guidelines [4] in combination with [3] (presented in subSections 2.2 and 2.3). Concerning this work, a careful manual screening of all articles is performed accordingly.

- *The research questions*

  Considering the different views on the research topic, we defined the following research questions:

  - *RQ 1:* Does the development of a DS(M)L follow a defined process? Can it be recognized as the utilization of the specified process?
  - *RQ 1.1:* Which parts of the process are used? More specifically, is it possible to recognize at least the main phases of the analysis and the design?
  - *RQ 2:* Which engineering principles are used while developing a DS(M)L?
  - *RQ 2.1:* How important are the agile principles in the development of a DS(M)L?
  - *RQ 3:* What is the role of the end user in the development of a DS(M)L?
  - *RQ 4:* Is the DS(M)L development actually supported by a specific tool?
  - *RQ 4.1:* Which kind of tool was developed to support DS(M)L use?

- *The collection sources*

  The articles are collected from three different types of sources:

  - The primary sources of articles were four digital libraries: Science Direct,[k] IEEE eXplore,[l] ACM DL,[m] and Web of Science.[n]
  - Next, several hundred articles were reused from the repository of a finalized SMS on DSL [9].
  - Additionally, we identified some excellent articles and performed (forward and backward) snowballing on them.
  - Finally, nonsystematic manual search added some more articles, too.

- *The inclusion/exclusion criteria*

  In short, we search for all articles in the English language, which were published between 2006 and 2016 and where a full PDF version of the article is available. We include the articles from journals and conference proceedings only. To exclude obviously inappropriate articles, we limit

---

[k] https://www.sciencedirect.com/.

[l] http://ieeexplore.ieee.org/.

[m] https://dl.acm.org/.

[n] http://login.webofknowledge.com.

the research field to Computer Science (if it was possible). To collect more evidence, we search for articles related to the development (process, approach) of a DSL or DSML in a less strict mode. The search string for the DL searching has the following form: DL AND (PR OR AD), where:

- DL = (model-driven engineering OR domain-specific language OR domain-specific modeling language OR MDE OR DSL OR DSML)
- PR = (process OR approach OR development)
- AD = (analysis AND design)

Table 7 presents the outcome of article collection. We collected 1541 articles in total from all available sources. After excluding the formally wrong articles (for instance, the duplicates and the articles with unavailable full contents) exactly 1350 articles remained.

- *The screening*

At first, the process of screening was planned to be performed manually and conducted in two phases. The first phase was conducted between May and November 2017. It included the screening of the title and abstract to eliminate the obviously wrong articles. The articles were classified into the three subsets:

- *the included* containing all suitable articles,
- *the possibly included* with the margin articles, which have to be read entirely before the final decision is taken, and
- *the excluded* containing unsuitable articles, thus the articles with unobtainable content on the right research topic, and the articles from other research areas and topics.

Table 7 The article collection for the SMS on DSL processes.

Source	Type	All found	Excluded	Used
Science direct	Digital library	93	19	74
IEEE eXplore	Digital library	365	47	318
ACM DL	Digital library	382	42	340
Web of science	Digital library	408	67	341
SMS on DSL's [9]	SMS research	256	15	241
Snowballing	Articles reference	24	1	23
Other	Manual search	13	0	13
Total		1541	191	1350

Table 8 The outcome of the manual screening.

	Included	Possibly included	Excluded
Number	375	266	699
Percentage	27.78%	19.70%	51.77%

Only the first two subsets of articles are used in the second phase of the screening. Here the full text of each article will be read to obtain the final decision to include or exclude the article.

By following the guidelines, we are currently performing the second phase of the screening and the final selection of articles is not over. Still, the results of the finalized first phase suffice for the scope of this work. The preliminary outcome of the manual screening is presented in Table 8. More than half of screened articles were excluded and less then a third included. The results obtained enable a general approximation of the screening expectations.

- *The classification*

A specific form is prepared to support the (manual) articles classification in relation to specified research questions. The detailed presentation is beyond the scope of this article. Similarly, other parts of the SMS protocol are omitted as well.

5.2 Conducting an experiment on automatic screening

In SE, a controlled experiment is defined as a randomized experiment or quasi-experiment in which individuals or teams (the experimental units) conduct one or more SE tasks for the sake of comparing different processes, methods, techniques, languages or tools (the treatments) [16, 55].

In our case, a quasi-experiment is performed to demonstrate the ability of the described automatic screening approach (and the tool) to adequately perform the required screening task. The general characteristics of our experiment are the following:

- *The input*: A set of 1350 articles was collected and prepared for the experiment. The JabRef° reference manager was used for organizing the collected materials. All articles were manually screened and the screening decisions saved as reference results in the appropriate mode (i.e., as the included, the possibly included and the excluded sets of articles).

° http://www.jabref.org.

- *The approach*: The automatic screening was performed several times to demonstrate several specific characteristics of the proposed solution. The execution strictly followed the approach presented in Section 4.1 and the tool presented in Section 4.2 was used to perform the core activities. Some manual manipulation was accordingly needed as well. All the results (i.e., the automatic screening decisions) were carefully saved along with the screening configuration information.

- *The experiment setting*: Two sets of articles were used in the experiment: a *SMALL* set with 76 articles (randomly selected from all articles), and the *BIG* set containing all 1350 articles. The first set was used to quickly test the experimental design. The second set provides more precise results and enables a better understanding of the size impact on the screening performance.

- *Experiment execution*: To raise the quality of conclusions, each experiment was performed in an iterative way using different values of the specific characteristic. To obtain more objective (i.e., representative) results, the principles of random article selection and execution repetition (e.g., five times) were strictly used. The final result is always an average of the results from all repetitions.

- *Improvement loop:* An important characteristic of our approach is the possibility of the iterative improvement of the decision rules. The experiment can use adaptive upgrading of the decision rules by applying different principles presented in Section 3.2.7. Different strategies (i.e., none, delicate, reasonable, strong, radical and brutal) are usually applied to better understand the consequences of their use.

- *Output*: As we measure conformance with the referential (i.e., manual) screening, two results were obtained. The *average mark* serves as a measure of the discrepancy between manual and automatic decisions, while the *percentage of the decisions taken* presents the practical usability of the result. Considering the optimal result equals the referential screening result and the referential result contains some "possibly included" decisions, the average mark is the low value, but rarely 0. Similarly, the percentage of the decision taken is not 100%, but equal to the referential percentage of the decisions taken.

The aim of our experiment is to demonstrate the usability of presented automatic screening solution as a possible replacement for manual screening. Most importantly, it must be capable (a) to achieve a suitable level of conformance with the referential results and (b) to perform the screening efficiently, thus with less manual effort.

To prove both goals, four issues were tested in the experiment:

- *The (optimal) size of pilot screening*: how many articles should be manually screened to define efficient decision rules?
- *The improvement strategy*: what strategy of decision rules improvement is the most adequate one? What is the effect of using more radical approaches?
- *The number and the type of different keywords*: how many keywords are needed for a delicate enough decision? What is the effect of positive and negative keywords on the efficiency of decision rules?
- *The grouping of keywords*: does the complexity of the decision rules structure and the use of logical expressions increase the quality of decisions taken?

All issues were investigated separately in four parts of the experiment. For each part the goal, experiment operation, precise settings and the main results are presented. The most important part is a concluding short discussion.

5.3 The size of pilot screening set

The idea behind this part of the experiment is to recognize the influence of the size of the pilot set on the quality of decisions. Obviously, a very small pilot set is not representative enough to enable an effective screening, while a large one may demand too much (i.e., more than necessary) manual effort.

- *Goal:* to recognize the most appropriate size of the manually screened pilot set using a fixed structure of the decision rules and an average decision improvement strategy.
- *Operation:* several different sizes of randomized pilot sets are used to define as good as a possible decision. To correctly check this issue, a simple decision rule structure with a fixed small number of keywords (11) is used. The decision rules are improved with the same (i.e., reasonable) improvement strategy (see Section 3.2.7).
- *Setting:* the experiment is performed on the SMALL ($n = 76$) and the BIG ($n = 1350$) set. In the first case, the sizes of the pilot set are adjusted on the values between 1 and 76 (increasingly by 5). In the second case, the sizes are set on the values between 25 and 1345 (increasing by 20).
- *Result: the average mark* and *the percentage of decisions taken* yield the specific automatic screening characteristics. The optimal result (i.e., manual screening result) is given for comparison on all graphs. The graphs for the SMALL set are presented in Figs. 4 and 5. Figs. 6 and 7 demonstrate the results for the BIG set, respectively.

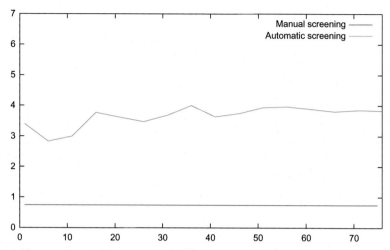

Fig. 4 The average mark using several different sizes of pilot sets ($n = 76$, reasonable improvement strategy).

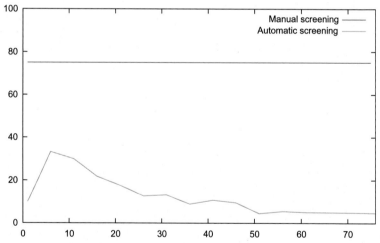

Fig. 5 The percentage of decisions taken using different sizes of pilot sets ($n = 76$, reasonable improvement strategy).

- *Discussion*: The results obtained (i.e., average mark around 4 and a very low percentage of the decisions taken in both cases) are very average, but this is caused by the specific decision rules structure and improvement strategy used. Our focus is on the difference between the results obtained for different sizes of pilot sets.

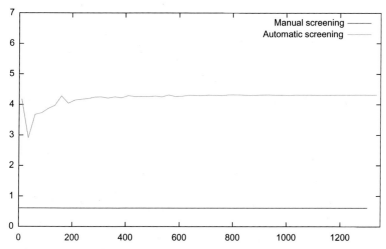

Fig. 6 The average mark using several different sizes of pilot sets ($n = 1350$, reasonable improvement strategy).

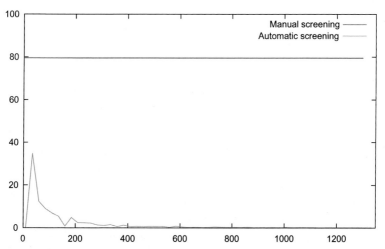

Fig. 7 The percentage of decisions taken using different sizes of pilot sets ($n = 1350$, reasonable improvement strategy).

Two important conclusions may be seen in the presented results. First, with increasing size of the pilot set both results to strive for a certain value, which seems to be specific for a fixed configuration of the decision rules. This is true for both experiment sets, but more clearly for the bigger set.

Second, even a relatively small pilot group enables a quite correct result. The margin size of the pilot set (i.e., the size, which results in approximately the same result as the outcome for a maximum size pilot set) has two characteristics: it must be absolutely bigger than some minimum value and it is related to the size of the entire set. In case of SMALL set of articles (see Figs. 4 and 5) and this specific setting, the margin size of the pilot set is approximately 50, thus more or less 65% of entire size. In the BIG set of articles (see Figs. 6 and 7), it is around 200, which means only 15% of the number of all articles.

5.4 The decision improvement strategy

The improvement strategy of the decision rules has an important impact on the quality of final decision rules. A very prudent strategy seems a good solution, but it might be sensitive to out-liners, thus a bigger pilot set (with more possible out-liners) might not yield better result as intuitively expected. On the other hand, a too radical strategy may lead to oversimplified decision rules, which contain a notable portion of the wrong decisions.

- *Goal*: to recognize the most suitable decision improvement strategy in relation to the size of the pilot set using a fixed structure of the decision rules.
- *Operation*: different sizes of the pilot sets are used to define the decision rules using a fixed simple structure with 11 keywords. The decisions are iteratively upgraded (five times) using different improvement strategies from the most careful (i.e., delicate) to the most radical one (see Section 3.2.7). The radical strategies skip one or more consecutive out-liners while defining the decision rules. The number of out-liners skipped is presented next to the strategy name (e.g., strong 1 strategy).
- *Setting*: similarly as before, the experiment is performed on the SMALL ($n = 76$) and the BIG ($n = 1350$) set using the same different sizes of the pilot set as in the first part of the experiment.
- *Result*: in Figs. 8 and 9, *the average mark* and *the percentage of decisions taken* in relation to the size of the pilot set are presented for different improvement strategies for the SMALL set of articles. Figs. 10 and 11 demonstrate the same characteristics for the BIG set.
- *Discussion:* The obvious conclusion is that applying a suitable improvement strategy is beneficial as it lowers the average mark and raises the percentage of decisions taken. Figs. 8 and 10 demonstrate the average

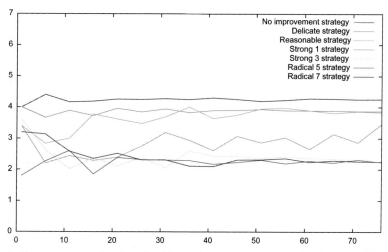

Fig. 8 The average mark using six different improvement strategies ($n = 76$, different sizes of pilot sets).

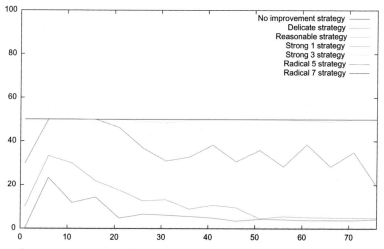

Fig. 9 The percentage of decisions taken using six different improvement strategies ($n = 76$, different sizes of pilot sets).

mark can be dropped from 4 to less than 2.3, which is a very good result for a small set of keywords.

Additionally, the use of more decisive (i.e., radical) improvement strategy yields better results. Such a strategy implies a bigger number of consciously wrong decisions (i.e., opposite to those of manual screening) to

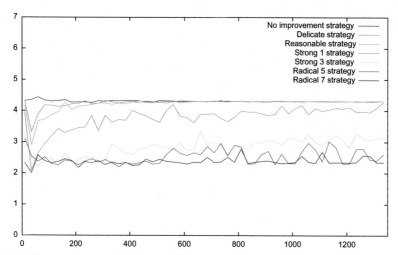

Fig. 10 The average mark using six different improvement strategies ($n = 1350$, different sizes of pilot sets).

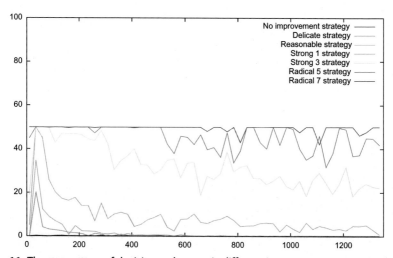

Fig. 11 The percentage of decisions taken us six different improvement strategies ($n = 1350$, different sizes of pilot sets).

gain a better overall score. To apply effective decisions adaptation, at least strong or radical strategies have to be implemented. Without this, the part of decisions taken is insignificant, too.

The average mark can be lowered down to a certain margin, which is specific for the screening configuration. Similarly, the percentage of decisions taken can rise up to 50%[P] (see Figs. 9 and 11). At this point, no additional profit is gained by using an even more radical strategy. Sometimes, the use of any improvement strategy which allows wrong decisions may be nonadmissible. In this case, a delicate or reasonable improvement strategy must be used.

With larger pilot set a more radical decision improvement is needed. In this way, the lack of decision content (originated in a too small set of keywords) is somehow compensated. In Figs. 10 and 11 the effect of very radical (i.e., brutal) improvement strategy is added to demonstrate this principle.

5.5 The number and type of different keywords

This part of experiment addresses an important issue in the definition of the decision rules: how many keywords should be used. Obviously, a (too) small set of keywords implies poor decision rules, but a large one might cause a complicated decision rule configuration and inefficient screening performance. Additionally, the introduction of negative keywords among the positive ones should enhance the quality of the final outcome, but it can prove problematic in case too general keywords are selected.

- *Goal*: to investigate the impact of using different sizes of (positive) keywords set, and the effect of the negative keywords addition.
- *Operation*: three different configurations of decision rules (i.e., one group with a different number of keywords) are used to screen the article set. The decision rules are five times iteratively upgraded using four improvement strategies (i.e., delicate, reasonable, strong, and brutal) on a fixed size pilot set of randomly selected articles.
- *Setting*: the first configuration (option 1 on Figs. 12–15) represents a *simple* solution, thus it includes a selection of 11 positive keywords only. The second configuration (option 2 on the figures) is an *extensive* one as it has a lot of positive keywords (62), while the third configuration (option 3 on the figures) is *extensive with negatives* as it includes 72 (positive and negative) keywords. The experiment is performed on the SMALL set of articles using the pilot set of 20 randomly selected articles. On the BIG set, the pilot set has fixed size 135.
- *Result*: Similarly as in other parts of experiment, *the average mark* and *the percentage of decisions taken* are presented in Figs. 12–15 for the SMALL and the BIG set, respectively.

[P] The value 50% is a consequence of the specific decision process implementation (see Section 3.2.5).

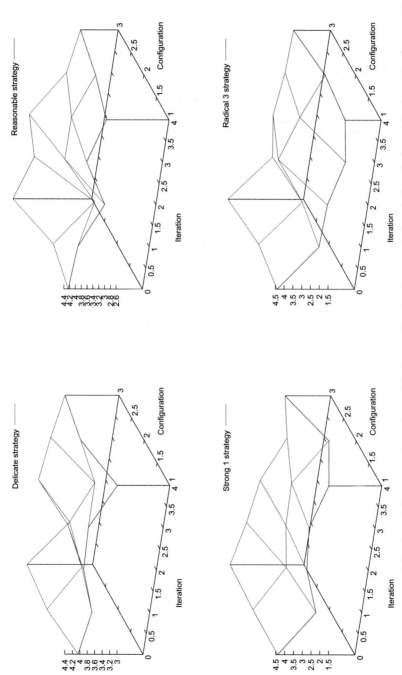

Fig. 12 The average mark for three different configurations of decision rules: simple, extended, and extended negative ($n = 76$, pilot $n = 20$, four different improvement strategies).

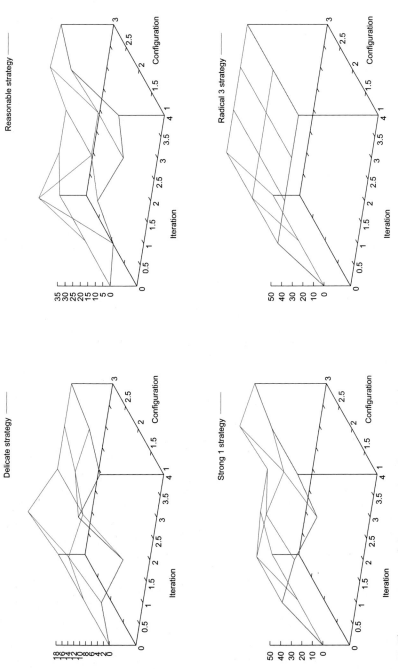

Fig. 13 The percentage of decisions taken for three different configurations of decision rules: simple, extended, and extended negative ($n = 76$, pilot $n = 20$, four different improvement strategies).

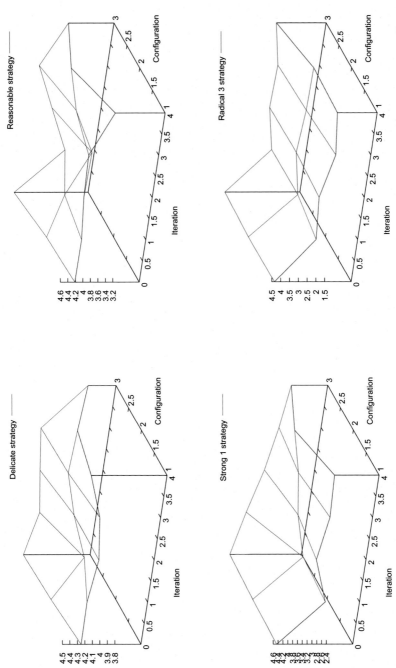

Fig. 14 The average mark for three different configurations of decision rules: simple, extended, and extended negative ($n = 1350$, pilot $n = 135$, four different improvement strategies).

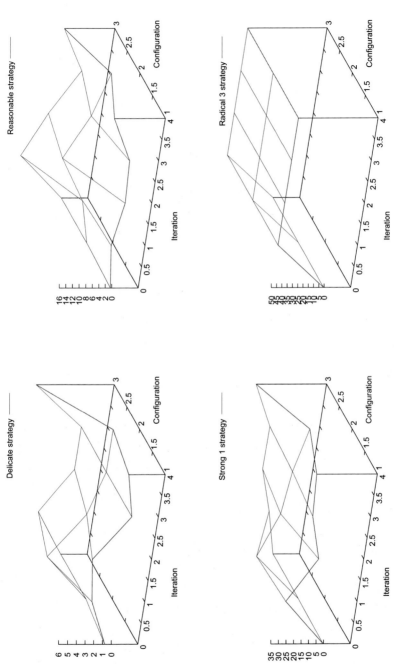

Fig. 15 The percentage of decisions taken for three different configurations of decision rules: simple, extended, and extended negative ($n = 1350$, pilot $n = 135$, four different improvement strategies).

- *Discussion*: Due to randomization and a moderate number of experiment repetitions the results are not aligned. Still, the results are clearly improved by iteratively applying selected improvement strategy. Also, both configurations with extensive keyword set yield slightly better results especially when we consider the percentage of the decisions taken. These observations are better visible for the bigger set of articles and more radical improvement strategies.

  On the other hand, the larger set of keywords does not automatically yield a much better result as intuitively expected. Thus, it is very dependent on keyword selection and demands a clear and operational definition of inclusion/exclusion rules in the previous phases of the screening process.

  In our case, the smallest set contains eleven carefully selected keywords, which prove crucial for the topic in question. In both extended sets, several synonyms and related positive keywords are added, but none of them was importantly different to significantly raise the quality of the final outcome. The same is true also for extended with negatives set of keywords.

  Still, introducing additional keywords to a specific keyword set is beneficial. It should at least retain the same result and sometimes automatically upgrade the screening performance, too. As our tool enables quick analysis, it enables iterative additions and consequently eases the task of keyword selection to the reviewer.

  The use of negative keywords is actually the same story as they must be very carefully chosen by a skilled reviewer. Due to our principle of the decision improvement, such words are quickly recognized as important ones for the exclusion. Thus, a more notable effect of the addition of only 11 negative keywords may be recognized (especially in Figs. 13 and 15).

  The results for small and big testing sets are very alike and compatible with previous results: the bigger set achieves a better average mark and demonstrates a stronger impact when using more radical decision improvement approach.

5.6 The structure of decision rules

The previous part of the experiment proves using an extensive keyword set does not automatically improve the solution. This is partly due to the difficult selection of correct keywords, but the structure of the decision rules (i.e., the distribution of keywords into the groups and the application of the group decisions principles) might have an impact, too. The last part of our experiment investigates this issue.

- *Goal*: to investigate the impact of keyword grouping (e.g., using a small or larger set of groups) and logical expression complexity.
- *Operation*: six different configurations of decision rules are applied on a fixed randomized pilot set and iteratively upgraded five times using the same four improvement strategies as before: delicate, reasonable, strong, and radical.
- *Setting*: the experiment is performed on the SMALL and the BIG set using the pilot set of 20 and 135 articles respectively. Six specific group configurations[q] of decision rules are investigated:
  - (1) includes 11 positive keywords grouped in 3 groups, the logical expression is simple;
  - (2) includes 11 positive keywords grouped in 3 groups, the logical expression is average;
  - (3) includes 62 positive keywords grouped in 11 groups, the logical expression is simple;
  - (4) includes 62 positive keywords grouped in 11 groups, the logical expression is complex;
  - (5) includes 71 positive and negative keywords grouped in 11 positive and one negative group, the logical expression is average;
  - (6) includes 71 positive and negative keywords grouped in 11 positive and one negative group, the logical expression is complex.
- *Result*: as before, *the average mark* and *the percentage of decisions taken* for both article sets are presented in four Figs. 16–19.
- *Discussion*: Due to similar experiment setting as in previous parts, the same observations can be seen in results of this experiment part. There are four new interesting observations.

First, the use of groups is beneficial and yields better results. It enables the definition of more elaborated rules, which are more adequate to implement inclusion/exclusion criteria by the reviewer. The same is observed in the small and in the big experiment.

Next, the intuition suggests the use of a more complex logical expression among groups should enable even better suitability. This is true for manual use, but due to our results not for automatic decision improvements. The principles of decision improvement are based on the

---

[q] The number in front of configuration description is used in Figs. 16–19 to identify specific configuration.

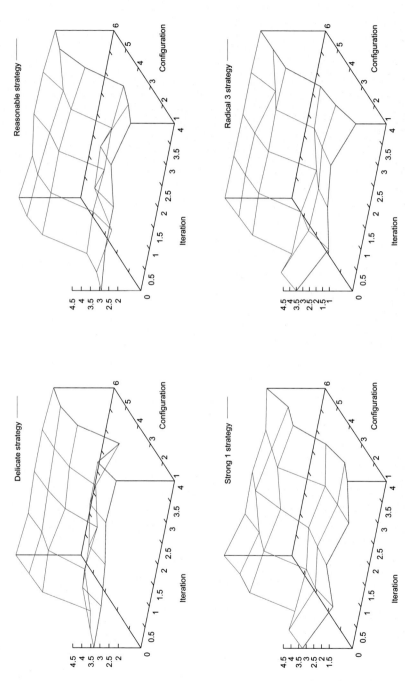

Fig. 16 The average mark for six different configurations of decision rules: simple, extended, and extended negative ($n = 76$, pilot $n = 20$, four different improvement strategies).

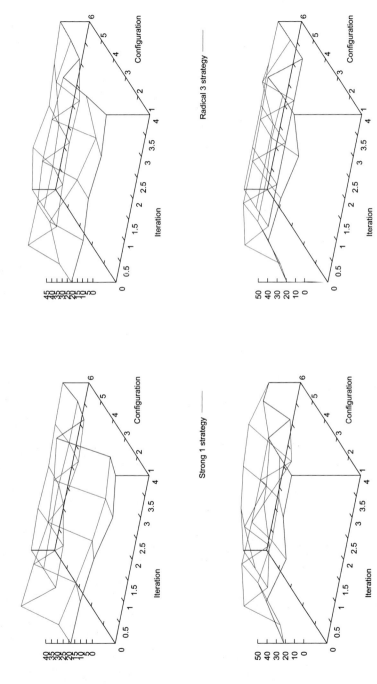

Fig. 17 The percentage of decisions taken for six different configurations of decision rules: simple, extended, and extended negative ($n = 76$, pilot $n = 20$, four different improvement strategies).

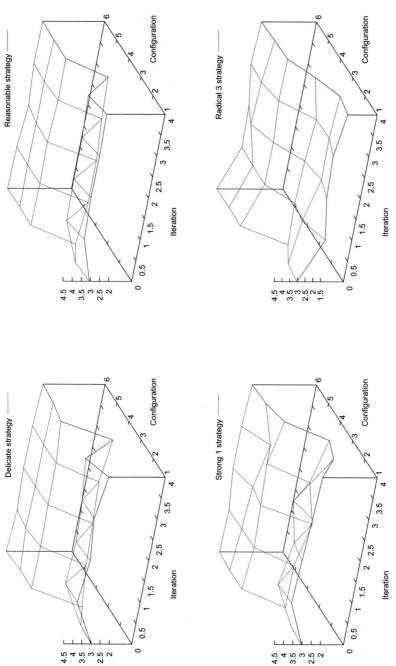

Fig. 18 The average mark for six different configurations of decision rules: simple, extended, and extended negative ($n = 1350$, pilot $n = 135$, four different improvement strategies).

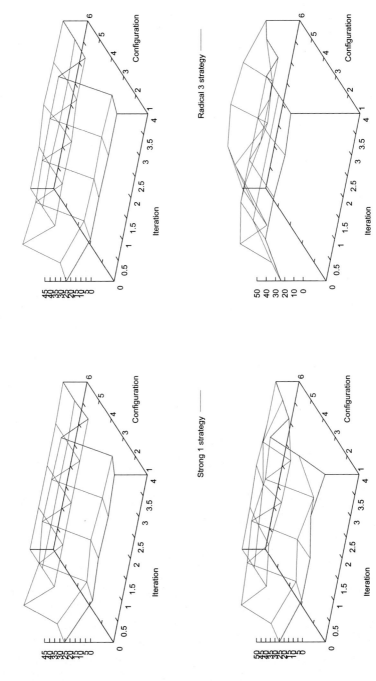

Fig. 19 The percentage of decisions taken for six different configurations of decision rules ($n = 1350$, pilot $n = 135$, four different improvement strategies).

principle of equality among keywords and groups. When the decisions are calculated differently (i.e., by logical expression), the solution does not work as expected. Consequently, the use of (simple) expressions enables is enabled as it is sometimes obligatory, but the use of the elaborate logical expressions requires caution. This is true for all sizes and improvement strategies.

Third, the configuration 3 is the best solution in our case as its average mark reaches down to less than 2 with the quite high percentage of decisions taken. This configuration includes a lot of positive keywords and groups, but only a simple conjunction of group decisions as a logical expression. In this case, it becomes possible to actually profit from the computing power, which is able to check a large number of possibilities.

Finally, not even the use of negative expressions can eliminate the effect of a (too) complex logical expression. The results are worse than in case of a smaller set of keywords. The results in Figs. 17 and 19 show even lower percentage of decisions taken.

The most satisfying result is demonstrated in Figs. 18 and 19 concerning the third configuration: an excellent result (i.e., a low average mark and a high percentage of the decisions taken) might be earned with the use of a careful strategy (e.g., delicate or reasonable), which does not imply risky steps (i.e., eliminate articles on purpose) in order to gain a better overall mark.

5.7 The concluding observations

The final step of the demonstration experiment is to generalize conclusions. The main observations may be summarized in five points:

- *The experiment proved the practical usability of our approach* Apart from presented results, the applicability of the presented approach and the tool's usability was tested by the experiment execution. By the experience, some improvements were introduced in the approach to fit better the actual performance. Once the initial issues with the tool use were over, the tool proved to be adaptable, easy to use and reliable.
- *The approach can be adapted to a variety of different research goals* Our experiment demonstrated the approach's applicability in case of conducting an SMS for a specific research topic. Due to the adaptable process and the decision rules configuration, the automatic screening can be applied to all types of research goals (e.g., SLR or SMS), different level of the screening accuracy, and certainly to a very broad variety of research topics from practically all research areas.

- *The crucial part is a detailed insight into research topic* The experiment proved the importance of previous steps of the SR research, especially the effective definition of inclusion/exclusion rules. The tool eases the configuration of the decision rules and it supports theirs iterative improvement in various ways, but the effective automatic screening grounds on clear understanding of the research topic (see Section 5.5).
- *Experienced reviewer is needed to efficiently perform automatic screening* Apart from clear insight into the research topic, the effective automatic screening requires a careful iterative decision rules adaptation performed by an experienced reviewer. The experiment proved the decision rules structure (i.e., the keywords, the groups, and the logical expression) empower the detailed enough implementation of the inclusion/exclusion criteria (see Section 5.6). The use of the most suitable improvement strategy (see Section 5.4) and the necessary number of the iterations are the reviewer's decision.
- *The approach and the supporting tool enable notable savings in the performance time* The main premise of our approach is to save the time using the automatic screening. The manual screening of the entire set of collected articles requires a considerable amount of time. Comparing to it, the time-consuming parts of the automatic screening are actually the time to configure effective decision rules and the manual screening of the pilot set articles. The first part may be variable, but an experienced reviewer using clear inclusion/exclusion criteria sets the configuration quickly. The second part is directly related to the size of the pilot set. Our experiment (see Section 5.3) demonstrated, the required number of articles to manually screen may be reduced considerably (e.g., to 15% in our realistic case) using our automatic screening approach.

6. Related work

Some authors proposed similar approaches to conduct an effective *selection of evidence for an SR using the text mining techniques.*
- In the work of Melheiros et al. [12], a visual text mining (VTM) approach was proposed to aid the SRs. The authors proposed a new strategy called VTM-Based Systematic Review and observed the beneficial results of its use.

    Our solution proposes a similar approach, but it is aimed to a specific part of SR process (i.e., to replace the manual screening part) rigorously following the guidelines. Additionally, our approach uses a different technique to gain the maximized impact.

- Felizardo et al. [11] proposed to use visual text mining approach called USR-VTM as an aid to support the selection of new primary studies when conducting an update of (an outdated) SR. They presented a tool named Revis and evaluated the approach through a comparison of results with the manual approach. The results proved the increased number of correctly included studies in comparison to the traditional approach. Consequently, USR-VTM effectively supports the update of an SRs.

  Apart from similarities, our approach aims to support the effective screening by using a statistical analysis approach, which can be adapted to the research topic to a chosen level, which suits best to the specific reviewer. This yields an expected effect and in indirectly regulates the amount of manual work required.

Additionally, some authors proposed to use the AI and *other advanced approaches to gain knowledge* from primary studies when conducting an SR.

- Matwin [32] used a simple and easy to implement a factorized version of the complement naive Bayes (FCNB) classifier to reduce the time spent by experts reviewing journal articles. The results recognized the usefulness of this approach for automation of systematic reviews of the drug class efficacy for the disease treatment.
- In the work of Cohen [33], the objective was to determine whether the automated classification of document citations can be useful in reducing the time spent by experts reviewing journal articles for inclusion in updating systematic reviews of drug class efficacy for treatment of disease.
- Ji et al. [34] demonstrated that utilizing established (lexical) article relationships and ontology-based semantics can facilitate the identification of relevant articles in an efficient and effective manner.
- Amarjeet [56] presented Fuzzy-Pareto dominance driven artificial bee colony (FP-ABC) algorithm for many-objective software module clustering, which can be used in article clustering, too.

Our solution uses a simpler and more understandable principle with the aim to facilitate the work of reviewers. Instead of using machine learning or ontology-based principles, the objective is to ease the actual performance of reviewers work even if they are not very familiar with the use of advanced computer techniques and tools. Still, our approach proposes an automatic improvement of the decision rules, which eases the reviewer's work and enables to configure the tool to the desired level.

Instead of performing an effective (automatic) screening over a large number of articles, the initial *number of articles might be limited* using an alternative approach.

- Kosar [7] proposed to limit the initial collection of articles when the number of collected articles guarantees that the error (because of possibly missing articles) falls under 5% margin. Consequently, this approach considerably speeds up the screening as well and can produce as much as 47% of savings in time.

  This approach might be used only in the case when the number of collected articles is very high. Thus, it is aimed at the SMSs only. Our solution may be used in case of an arbitrary initial number of articles, thus it is applicable for conducting SLRs as well.

7. Conclusion and future work

7.1 Conclusion

The aim of our work is to propose, implement, and demonstrate a solution which speeds up the screening phase of the SR process. The main idea is to appropriately substitute manually performed screening with a screening, which is (at least) partly performed automatically while carefully assesses the quality of the outcome. The presented approach is described in detail and integrated into the rigorous SLR or SMS process. It is implemented by an easy-to-use tool, which is able to organize and efficiently perform the screening. The final demonstration experiment on the currently conducted SMS proved the practical value of the described approach.

A few important benefits should be emphasized at this point. First, the approach is rigorously *designed to be consistent with the strict SR guidelines* where the extensive SR experience provided in last decades is summarized. In this way, the correct performance and generality are achieved.

Second, the approach is *operational in practice* as it takes into account some specifics and issues from the practical screening execution (e.g., the article format transformation). Our experience in conducting the SMSs earned valuable insight into the screening specifics and solution possibilities. The main consequences are reflected in the tool which implements an adjustable combination of the manual and the automatic screening, and the expected active involvement of an experienced screening reviewer.

Third, the approach *defines a highly adjustable screening process* which is able to comply with very different screening situations. The iterative nature of the decision rules definition can be set to sharply decide without any manual work or to screen the entire set of collected articles. Usually, at least a small pilot set has to be screened manually to configure the decision rules.

Fourth, the approach defines an efficient solution as it *implements a specific combination of carefully selected principles*. The implementation of the decision rules by the text statistic analysis empowers an understandable configuration by the screening performers. In combination with the iterative improvement based on the marking of the decisions provides a powerful enhancement of the solution. An important aim of the tool is to manage such decision rules configuration thus providing to reviewer an efficient support to successfully conduct the screening.

Finally and most importantly, *the approach yields successful in practice*. Due to versatile nature of the articles in SE, the optimal result (i.e., total match with manual screening) may be unreachable in practice, but good results can be efficiently obtained by proper decision configuration setting. Our experiment proved the size of the pilot set can be fairly small (in our case 5% of all articles). Obviously, the efficient screening configuration bases on clear inclusion/exclusion criteria. In this case, two or three iterative configuration updates using a suitable improvements strategy yield very good results already. Thus, considerable savings in time are actually achieved.

7.2 Future work

Currently, the approach is practically tested on the SMS research on examining different processes while developing a specific DSL or DSML. Apart from testing tools quality, such research task induces new ideas for future work, too.

The first task is to perform additional and more profound testings of the proposed solution. Apart SMS, they should include the screening of articles in the SLR research. Performing the research on different research topics and in diverse research areas might yield new issues to solve, too.

Further, an upgrade of the existing tool functionality might be needed as it can be still considered the first version of solution implementation. These enhancements are not aimed toward better performance and usability only. They should introduce new advanced solutions such as an implementation of AI principles in the core of decision-making and in the decision rules improvement strategies.

References

[1] B.A. Kitchenham, T. Dyba, M. Jorgensen, Evidence-based software engineering, in: 26th International Conference on Software Engineering (ICSE), 2004, pp. 273–281, ISBN: 0-7695-2163-0, https://doi.org/10.1109/ICSE.2004.1317449.
[2] B.A. Kitchenham, Procedures for Performing Systematic Reviews (TR/SE-0401), vol. 33, Keele University, Keele, UK, 2004, p. 28. 10.1.1.122.3308.

[3] K. Petersen, S. Vakkalanka, L. Kuzniarz, Guidelines for conducting systematic mapping studies in software engineering: an update, Inf. Softw. Technol. 64 (2015) 1–18, https://doi.org/10.1016/j.infsof.2015.03.007.

[4] B.A. Kitchenham, S. Charters, Guidelines for performing systematic literature reviews in software engineering version 2.3, Engineering 45 (4ve) (2007) 1051, https://doi.org/10.1145/1134285.1134500.

[5] C. Wohlin, P. Runeson, P.A. Da Mota Silveira Neto, E. Engström, I. Do Carmo Machado, E.S. De Almeida, On the reliability of mapping studies in software engineering, J. Syst. Softw. 86 (10) (2013) 2594–2610, https://doi.org/10.1016/j.jss.2013.04.076.

[6] P. Brereton, B.A. Kitchenham, D. Budgen, M. Turner, M. Khalil, Lessons from applying the systematic literature review process within the software engineering domain, J. Syst. Softw. 80 (4) (2007) 571–583, https://doi.org/10.1016/j.jss.2006.07.009.

[7] T. Kosar, S. Bohra, M. Mernik, A systematic mapping study driven by the margin of error, J. Syst. Softw. 144 (2018) 439–449, https://doi.org/10.1016/j.jss.2018.06.078.

[8] E. Syriani, L. Luhunu, H. Sahraoui, Systematic mapping study of template-based code generation, Comput. Lang. Syst. Struct. 52 (2018) 43–62, https://doi.org/10.1016/j.cl.2017.11.003.

[9] T. Kosar, S. Bohra, M. Mernik, Domain-specific languages: a systematic mapping study, Inf. Softw. Technol. 71 (2016) 77–91, https://doi.org/10.1016/j.infsof.2015.11.001.

[10] F. Febrero, C. Calero, M. Ángeles Moraga, A systematic mapping study of software reliability modeling, Inf. Softw. Technol. 56 (8) (2014) 839–849, https://doi.org/10.1016/j.infsof.2014.03.006.

[11] K.R. Felizardo, E.Y. Nakagawa, S.G. MacDonell, J.C. Maldonado, A visual analysis approach to update systematic reviews, in: Proceedings of the 18th International Conference on Evaluation and Assessment in Software Engineering, ACM, New York, NY, USA, 2014, pp. 4:1–4:10, ISBN: 978-1-4503-2476-2, https://doi.org/10.1145/2601248.2601252.

[12] V. Malheiros, E. Hohn, R. Pinho, M. Mendonca, J.C. Maldonado, A visual text mining approach for systematic reviews, in: Proceedings of the First International Symposium on Empirical Software Engineering and Measurement, IEEE Computer Society, Washington, DC, USA, 2007, pp. 245–254, ISBN: 0-7695-2886-4, https://doi.org/10.1109/ESEM.2007.13.

[13] V. Blagojević, D. Bojić, M. Bojović, M. Cvetanović, J. Djordjević, D. Djurdjević, B. Furlan, S. Gajin, Z. Jovanović, D. Milićev, V. Milutinović, B. Nikolić, J. Protić, M. Punt, Z. Radivojević, ž. Stanisavljević, S. Stojanović, I. Tartalja, M. Tomašević, P. Vuletić, Chapter One—a systematic approach to generation of new ideas for PhD research in computing, in: A.R. Hurson, V. Milutinović (Eds.), Creativity in Computing and DataFlow SuperComputing, Advances in Computers, vol. 104, Elsevier, 2017, pp. 1–31, https://doi.org/10.1016/bs.adcom.2016.09.001.

[14] B. Spring, Evidence-based practice in clinical psychology: what it is, why it matters; what you need to know, J. Clin. Psychol. 63 (7) (2007) 611–631, https://doi.org/10.1002/jclp.

[15] M.V. Zelkowitz, D. Wallace, Experimental validation in software engineering, Inf. Softw. Technol. 39 (11) (1997) 735–743. 16/S0950-5849(97)00025-6.

[16] R. Wieringa, Empirical research methods for technology validation: scaling up to practice, J. Syst. Softw. 95 (2014) 19–31, https://doi.org/10.1016/j.jss.2013.11.1097.

[17] Centre for Reviews and Dissemination, Systematic Reviews: CRD's Guidance for Undertaking Reviews in Health Care, University of York, 2009, p. 294, ISBN: 1900640473, https://doi.org/10.1016/S1473-3099(10)70065-7.

[18] T. Dybå, T. Dingsøyr, Strength of evidence in systematic reviews in software engineering, in: Proceedings of the Second ACM-IEEE International Symposium on Empirical

Software Engineering and Measurement, ACM, New York, NY, USA, 2008, pp. 178–187, ISBN: 978-1-59593-971-5, https://doi.org/10.1145/1414004.1414034.

[19] B.A. Kitchenham, S.L. Pfleeger, L.M. Pickard, P.W. Jones, D.C. Hoaglin, K. El Emam, J. Rosenberg, Preliminary guidelines for empirical research in software engineering, IEEE Trans. Softw. Eng. 28 (8) (2002) 721–734, https://doi.org/10.1109/TSE.2002.1027796.

[20] University of York Centre for Reviews and Dissemination, Undertaking systematic reviews of research on effectiveness: CRD's guidance for carrying out or commissioning reviews, in: K.S. Khan, G. ter Riet, J. Glanville, A.J. Sowden, J. Kleijnen (Eds.), Undertaking Systematic Reviews of Research on Effectiveness: CRD's Guidance for Those Carrying Out or Commissioning Reviews, second ed., NHS Centre for Reviews and Dissemination, York, UK, 2001. Centre for Reviews and Dissemination Report 4.

[21] M. Petticrew, H. Roberts, Systematic Reviews in the Social Sciences: A Practical Guide, Blackwell Publishing, 2005, ISBN: 1405121106.

[22] C. Wohlin, Guidelines for snowballing in systematic literature studies and a replication in software engineering, in: Proceedings of the 18th International Conference on Evaluation and Assessment in Software Engineering, ACM, New York, NY, USA, 2014, pp. 38:1–38:10, ISBN: 978-1-4503-2476-2, https://doi.org/10.1145/2601248.2601268.

[23] K. Petersen, R. Feldt, S. Mujtaba, M. Mattsson, Systematic mapping studies in software engineering, in: 12th International Conference on Evaluation and Assessment in Software Engineering, vol. 17, 2008, p. 10, ISBN: 0-7695-2555-5, https://doi.org/10.1142/S0218194007003112.

[24] L.M. do Nascimento, D.L. Viana, P.A.M.S. Neto, D.A.O. Martins, V.C. Garcia, S.R.L. Meira, A systematic mapping study on domain-specific languages, in: Seventh International Conference on Software Engineering and Advances (ICSEA 2012), 2012, pp. 179–187, ISBN: 9781612082301.

[25] A. Mehmood, D.N.A. Jawawi, Aspect-oriented model-driven code generation: a systematic mapping study, Inf. Softw. Technol. 55 (2) (2013) 395–411, https://doi.org/10.1016/j.infsof.2012.09.003.

[26] V. Garousi, A. Mesbah, A. Betin-Can, S. Mirshokraie, A systematic mapping study of web application testing, Inf. Softw. Technol. 55 (8) (2013) 1374–1396, https://doi.org/10.1016/j.infsof.2013.02.006.

[27] E. Engström, P. Runeson, Software product line testing—a systematic mapping study, Inf. Softw. Technol. 53 (1) (2011) 2–13, https://doi.org/10.1016/j.infsof.2010.05.011.

[28] D. Budgen, M. Turner, P. Brereton, B.A. Kitchenham, Using mapping studies in software engineering, in: Proceedings of the Psychology of Programming Interest Group, vol. 2, 2008, pp. 195–204.

[29] B.A. Kitchenham, D. Budgen, O. Pearl Brereton, Using mapping studies as the basis for further research: a participant-observer case study, Inf. Softw. Technol. 53 (6) (2011) 638–651, https://doi.org/10.1016/j.infsof.2010.12.011.

[30] H. Arksey, L. O'Malley, Scoping studies: towards a methodological framework, Int. J. Social Res. Methodol. 8 (1) (2005) 19–32, https://doi.org/10.1080/1364557032000119616.

[31] B.A. Kitchenham, O. Pearl Brereton, D. Budgen, M. Turner, J. Bailey, S. Linkman, Systematic literature reviews in software engineering—a systematic literature review, Inf. Softw. Technol. 51 (1) (2009) 7–15, https://doi.org/10.1016/j.infsof.2008.09.009.

[32] S. Matwin, A. Kouznetsov, D. Inkpen, O. Frunza, P. O'Blenis, A new algorithm for reducing the workload of experts in performing systematic reviews, J. Am. Med. Informatics Assoc. 17 (4) (2010) 446–453, https://doi.org/10.1136/jamia.2010.004325.

[33] A.M. Cohen, W.R. Hersh, K. Peterson, P.Y. Yen, Reducing workload in systematic review preparation using automated citation classification, J. Am. Med. Inform. Assoc. 13 (2) (2006) 206–219, https://doi.org/10.1197/jamia.M1929.

[34] X. Ji, A. Ritter, P.Y. Yen, Using ontology-based semantic similarity to facilitate the article screening process for systematic reviews, J. Biomed. Informatics 69 (2017) 33–42, https://doi.org/10.1016/j.jbi.2017.03.007.

[35] S.J. Mellor, K. Scott, A. Uhl, D. Weise, MDA Distilled—Principles of Model-Driven Architecture, Addison-Wesley, Boston, MA, USA, 2004.

[36] I. Rožanc, B. Slivnik, Using reverse engineering to construct the platform independent model of a web application for student information systems, Comput. Sci. Inf. Syst. 10 (4) (2013) 1557–1583.

[37] T. Kosar, N. Oliveira, M. Mernik, M.J. Pereira Varanda, M. Črepinsek, D. da Cruz, P.R. Henriques, Comparing general-purpose and domain-specific languages: an empirical study, Comput. Sci. Inform. Syst. 7 (2) (2010) 247–264.

[38] M. Mernik, B.R. Bryant, Special issue on the programming Languages track at the 28th ACM symposium on applied computing, Comput. Lang. Syst. Struct. 40 (1) (2014) 1, https://doi.org/10.1016/j.cl.2014.03.002.

[39] P. Mohagheghi, V. Dehlen, N. Tor, Definitions and approaches to model quality in model-based software development—a review of literature, Inf. Softw. Technol. 51 (12) (2009) 1646–1669.

[40] M. Bastarrica, S. Rivas, P. Rossel, Designing and implementing a product family of model consistency checkers, in: Proceedings of the 10th ACM/IEEE International Conference on Model Driven Engineering Languages and Systems (MoDELS 2007)—Quality in Modeling Workshop, October, Nashville, TN, USA, 2007, pp. 36–49.

[41] B.R. Bryant, J.G. Gray, M. Mernik, P.J. Clarke, R.B. France, G. Karsai, Challenges and directions in formalizing the semantics of modeling languages, Comput. Sci. Inf. Syst. 8 (2011) 225–253.

[42] I. Fister, I. Jr, Fister, M. Mernik, J. Brest, Design and implementation of domain-specific language easytime, Comput. Lang. Syst. Struct. 37 (4) (2011) 151–167.

[43] I. Luković, M.J. Varanda Pereira, N. Oliveira, D. da Cruz, P.R. Henriques, A DSL for PIM specifications: design and attribute grammar based implementation, Comput. Sci. Inf. Syst. 8 (2) (2011) 379–403.

[44] T. Kosar, M. Mernik, J. Gray, T. Kos, Debugging measurement systems using a domain-specific modeling language, Comput. Ind. 65 (4) (2014) 622–635, https://doi.org/10.1016/j.compind.2014.01.013.

[45] A. Barišić, V. Amaral, M. Goulão, Usability driven DSL development with USE-ME, Comput. Lang. Syst. Struct. 51 (Suppl. C) (2018) 118–157, https://doi.org/10.1016/j.cl.2017.06.005.

[46] I. Rožanc, B. Slivnik, On the appropriateness of domain-specific languages derived from different metamodels, in: 2014 9th International Conference on the Quality of Information and Communications Technology, September, 2014, pp. 190–195, https://doi.org/10.1109/QUATIC.2014.33.

[47] D. Méndez-Acuña, J.A. Galindo, T. Degueule, B. Combemale, B. Baudry, Leveraging software product lines engineering in the development of external DSLs: a systematic literature review, Comput. Lang. Syst. Struct. 46 (Suppl. C) (2016) 206–235, https://doi.org/10.1016/j.cl.2016.09.004.

[48] M. Mernik, J. Heering, A.M. Sloane, When and how to develop domain-specific languages, ACM Comput. Surv. 37 (4) (2005) 316–344.

[49] M. Strembeck, U. Zdun, An approach for the systematic development of domain-specific languages, Softw. Pract. Experience 39 (7) (2009) 1253–1292, https://doi.org/10.1002/spe.936.

[50] E. Visser, WebDSL: a case study in domain-specific language engineering, in: Generative and Transformational Techniques in Software Engineering II, vol. 5235, 2008, pp. 291–373.

[51] I. Sommerville, Software Engineering, ninth ed., Addison-Wesley Publishing Company, USA, 2010. 0137035152, 9780137035151.

[52] K. Beck, M. Beedle, A. van Bennekum, A. Cockburn, W. Cunningham, M. Fowler, J. Grenning, J. Highsmith, A. Hunt, R. Jeffries, J. Kern, B. Marick, R.C. Martin, S. Mellor, K. Schwaber, J. Sutherland, D. Thomas, Manifesto for Agile Software Development, 2001. Retrieved from http://www.agilemanifesto.org/ (December 29, 2017).

[53] V. Mahnič, A case study on agile estimating and planning using scrum, Electron. Electr. Eng. 5 (111) (2011) 123–128.

[54] M.J. Villanueva Del Pozo, An Agile Model-Driven Method for Involving End-Users in DSL Development (Ph.D. thesis), Universitat Politècnica de València, Valencia (Spain), 2016, 10.4995/Thesis/10251/60156.

[55] D.I.K. Sjoberg, J.E. Hannay, O. Hansen, V. By Kampenes, A. Karahasanovic, N.-K. Liborg, A.C. Rekdal, A survey of controlled experiments in software engineering, IEEE Trans. Softw. Eng. 31 (9) (2005) 733–753, https://doi.org/10.1109/TSE.2005.97.

[56] Amarjeet, J.K. Chhabra, FP-ABC: Fuzzy-Pareto dominance driven artificial bee colony algorithm for many-objective software module clustering, Comput. Lang. Syst. Struct. 51 (Suppl. C) (2018) 1–21, https://doi.org/10.1016/j.cl.2017.08.001.

About the authors

Igor Rožanc received the MSc and PhD degrees in Computer Science from the University of Ljubljana in 1996 and 2003, respectively. He is currently a senior lecturer at the University of Ljubljana, Faculty of Computer and Information Science. His research interests are in the area of Software Engineering, especially software quality, model-driven development, and domain-specific (modeling) languages. He is a member of the IEEE. More information about him is available at https://www.fri.uni-lj.si/en/employees/igor-rozanc.

Marjan Mernik received the MSc and PhD degrees in Computer Science from the University of Maribor in 1994 and 1998, respectively. He is currently a professor at the University of Maribor, Faculty of Electrical Engineering and Computer Science. He was a visiting professor at the University of Alabama at Birmingham, Department of Computer and Information Sciences. His research interests include programming languages, compilers, domain-specific (modeling) languages, grammar-based systems, grammatical inference, and evolutionary computations. He is a member of the IEEE, ACM, and EAPLS. He is the Editor-in-Chief of the Journal of Computer Languages, as well as Associate Editors of the Applied Soft Computing Journal, Information Sciences Journal, and Swarm and Evolutionary Computation Journal. He is being named a Highly Cited Researcher for years 2017 and 2018. More information about his work is available at https://lpm.feri.um.si/en/members/mernik/.

CHAPTER FOUR

A survey on cloud-based video streaming services

Xiangbo Li[a], Mahmoud Darwich[b], Mohsen Amini Salehi[c], and Magdy Bayoumi[d]

[a]Amazon AWS IVS, San Diego, California, CA, United States
[b]Department of Mathematical and Digital Sciences, Bloomsburg University of Pennsylvania, Bloomsburg, PA, United States
[c]School of Computing and Informatics, University of Louisiana at Lafayette, Lafayette, LA, United States
[d]Department of Electrical and Computer Engineering, University of Louisiana at Lafayette, Lafayette, LA, United States

Contents

Advances in Computers, Volume 123
ISSN 0065-2458
https://doi.org/10.1016/bs.adcom.2021.01.003

193

Abstract

Video streaming, in various forms of video on demand (VOD), live, and 360 degree streaming, has grown dramatically during the past few years. In comparison to traditional cable broadcasters whose contents can only be watched on TVs, video streaming is ubiquitous and viewers can flexibly watch the video contents on various devices, ranging from smartphones to laptops, and large TV screens. Such ubiquity and flexibility are enabled by interweaving multiple technologies, such as video compression, cloud computing, content delivery networks, and several other technologies. As video streaming gains more popularity and dominates the Internet traffic, it is essential to understand the way it operates and the interplay of different technologies involved in it. Accordingly, the first goal of this paper is to unveil sophisticated processes to deliver a raw captured video to viewers' devices. In particular, we elaborate on the video encoding, transcoding, packaging, encryption, and delivery processes. We survey recent efforts in academia and industry to enhance these processes. As video streaming industry is increasingly becoming reliant on cloud computing, the second goal of this survey is to explore and survey the ways cloud services are utilized to enable video streaming services. The third goal of the study is to position the undertaken research works in cloud-based video streaming and identify challenges that need to be obviated in future to advance cloud-based video streaming industry to a more flexible and user-centric service.

List of abbreviations

AVC	Advanced Video Coding
AWS	Amazon Web Services
CDN	content delivery networks
CMAF	Common Media Application Format
CPU	central processing unit
CVSE	cloud-based video streaming engine
DASH	Dynamic Adaptive Streaming over HTTP
DRM	digital rights management
DVD	digital versatile disc
GOP	group of pictures
HEVC	High Efficiency Video Coding
HLS	HTTP Live Streaming
HTTP	Hypertext Transfer Protocol
ISOBMFF	ISO base media file format
M2TS	MPEG-2 transport stream
M3U8	MP3 Playlist File (UTF-8)
MAC	message authentication code
MB	macroblock
MCU	multipoint control unit
MPD	media presentation description
MPEG	Moving Picture Experts Group
OTT	over the top
P2P	peer-to-peer

QoE	Quality of Experience
RTMP	Real-Time Messaging Protocol
RTP	Real-Time Protocol
RTSP	The Real-Time Streaming Protocol
SDTP	stall duration tail probability
SLA	service level agreement
SLO	service level objectives
UDP	User Datagram Protocol
VM	virtual machine
VOD	video on demand
VP	video phone
VRC	video cassette recording
XML	Extensible Markup Language

1. Introduction

1.1 Overview

The idea of receiving a stream of video contents dates back to the invention of television in the early years of the 20th century. However, the medium on which people receive and watch video contents has substantially changed during the past decade—from conventional televisions to streaming on a wide variety of devices (e.g., laptops, desktops, and tablets) via Internet. Adoption of the Internet-based video streaming is skyrocketing to the extent that it has dominated the whole Internet traffic. A report by Global Internet Phenomena shows that video streaming has already accounted for more than 60% of the whole Internet traffic [1]. The number of Netflix[a] subscribers has already surpassed cable-TV subscribers in the United States [2].

Nowadays, many Internet-based applications function based on video streaming. Such applications include user-generated video contents (e.g., those in YouTube,[b] Vimeo[c]), live streaming and personal broadcasting through social networks (e.g., UStream[d] and Facebook Live[e]), over the top (OTT) streaming (e.g., Netflix and Amazon Prime[f]), e-learning systems

---

[a] https://www.netflix.com.
[b] https://www.youtube.com.
[c] https://www.vimeo.com.
[d] https://video.ibm.com.
[e] https://www.facebook.com.
[f] https://www.amazon.com.

[3] (e.g., Udemy[g]), live game streaming platform (e.g., Twitch[h]), video chat and conferencing systems [4], natural disaster management and security systems that operate based on video surveillance [5], and network-based broadcasting channels (e.g., news and other TV channels) [6].

As video streaming services grow in popularity, they demand more computing services for streaming. The uprising popularity and adoption of streaming has coincided with the prevalence of cloud computing technology. Cloud providers offer a wide range of computing services and enable users to outsource their computing demands. Cloud providers relieve video streaming providers from the burden and implications of maintaining and upgrading expensive computing infrastructure [7]. Currently, video stream providers are extensively reliant on cloud services for most or all of their computing demands [8]. The marriage of video streaming and cloud services has given birth to a set of new challenges, techniques, and technologies in the computing industry.

Although numerous research works have been undertaken on cloud-based video streaming, to our knowledge, there is no comprehensive survey that shed lights on challenges, techniques, and technologies in cloud-based video streaming. As such, the essence of this study is to *first*, shed light on the sophisticated processes required for Internet-based video streaming; *second*, provide a holistic view on the ways cloud services can aid video stream providers; *third*, provide a comprehensive survey on the research studies that were undertaken in the intersection of video streaming and cloud computing; and *fourth* discuss the future of cloud-based video streaming technology and identify possible avenues that require further research efforts from industry and academia.

Accordingly, in this study, we first explain the way video streaming works and elaborate on each process involved in it. Then, in Section 3, we provide a holistic view of the challenges and demands of the current video streaming industry. Next, in Section 4, we discuss how cloud services can fulfill the demands of video streaming, and the survey the research works undertaken for that purpose. In the end, in Section 5, we discuss the emerging research areas in the intersection of video streaming and cloud computing.

1.2 Cloud computing for video streaming

To provide a high Quality of Experience (QoE) for numerous viewers scattered worldwide with diverse display devices and network characteristics,

---

[g] https://www.udemy.com.

[h] https://www.twitch.tv.

video stream providers commonly preprocess (e.g., pretranscode and prepackage) and store the video contents of multiple versions [9]. As such, viewers with different display devices can readily find the version matches their devices. An alternative approach to preprocessing videos is to process them in a lazy (i.e., on-demand) manner [7, 9]. In particular, this approach can be useful for videos that are rarely accessed. Recent analytical studies on the statistical patterns of accessing video streams (e.g., [10]) reveal that the videos of a repository are not uniformly accessed. In fact, streaming videos follows a long-tail pattern [11]. That means, many video streams are rarely accessed and only a small portion (approximately 5%) of the videos (generally referred to as *hot video streams*) are frequently accessed. Both preprocessing and on-demand approaches need the video streaming provider to provide an enormous computing facility.

Maintaining and upgrading in-house infrastructure to fulfill the computational, storage, and networking demands of video stream providers is costly. Besides, it is technically far from the mainstream business of stream providers, which is video content production and publishing. Alternatively, cloud service providers, such as Amazon Cloud (AWS), Google Cloud, and Microsoft Azure, can satisfy these demands by offering efficient services with a high availability [12]. Video stream providers have become extensively reliant on cloud services for most or all of their computing demands. For instance, Netflix has outsourced the entirety of its computational demands to Amazon cloud [13].

In spite of numerous advantages, deploying cloud services has presented new challenges to video stream providers. In particular, as cloud providers charge their users in a pay-as-you-go manner for their services [14], the challenge for stream providers is to minimize their cloud expenditure, while offering a certain level of QoE for their viewers.

Numerous research works have been conducted to overcome the challenges of video streaming using cloud services. For instance, researchers have studied the cost of deploying cloud services to transcode videos [12, 15], the cost-benefit of various video segmentation models [9, 16], applying customized scheduling methods for video streaming tasks [15, 17], and resource (virtual machine) provisioning methods for video streaming [9, 18]. Nonetheless, these studies mostly concentrate on one aspect of video streaming processing and how that aspect can be performed efficiently on the cloud. Further studies are required to position these works in a bigger picture and provide a higher level of vision on the efficient use of cloud services for the entire workflow of video streaming.

2. The mystery of video streaming operation

2.1 Structure of a video stream

As shown in Fig. 1, a video stream consists of a sequence of multiple smaller segments. Each segment contains several *group of pictures* (GOP) with a segment header at the beginning of each GOP. The segment header includes information such as the number of GOPs in that segment and the type of the GOPs. A GOP is composed of a sequence of frames. The first frame is an I (intra) frame, followed by several P (predicted) and B (bi-directional predicted) frames. A frame in a GOP is divided into multiple *slices* and each slice consists of several *macroblocks* (MB). The MBs are considered as the unit for video encoding and decoding operations.

Two types of GOP can exist, namely closed-GOP and open-GOP. In the former, GOPs are independent, i.e., there is no relation between

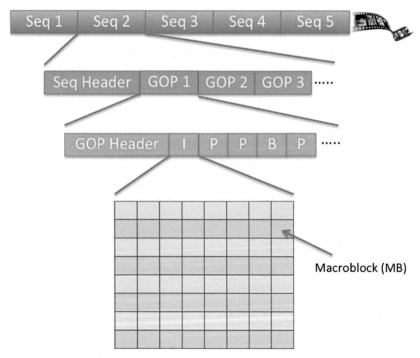

Fig. 1 The structural overview of a video stream. A video stream contains multiple segments and each segment contains multiple GOPs. A frame in a GOP includes a number of macroblocks (MB).

GOPs. Therefore, closed-GOPs can be processed independently. Alternatively, in open-GOP, a GOP is dependent on another GOP, hence, cannot be processed independently.

To process video streams, they can be split at different levels, namely segment level, GOP level, frame level, slice level, and MB level. Sequence-level contains several GOPs that can be processed independently. However, due to the large size of a sequence, its transmission and processing time become a bottleneck [19]. Processing at the frame, slice, and MB levels implies dealing with the spatio-temporal dependencies that makes the processing complicated and slow [19]. In practice, video stream providers generally perform processing at the segment or GOP levels. That is, they define a segment or a GOP as a unit of processing (i.e., a *task*) that can be processed independently [16].

2.2 Video streaming operation workflow

In both video on demand (VOD) (e.g., Hulu, YouTube, Netflix) and Live streaming (e.g., Livestream[i]), the video contents generated by cameras have to go through a complex workflow of processes before being played on viewers' devices. In this section, we describe these processes.

Fig. 2 provides a bird's-eye view of the main processes performed for streaming a video—from video production to playing the video on viewers' devices. These processes collectively enable raw and bulky videos, generated by cameras, be played on a wide variety of viewers' devices in a real-time manner and with the minimum delay. It is noteworthy that, in addition to these processes, there are generally other processes to enable features such as video content protection and cost-efficiency of video streaming. In the rest of this section, we elaborate on the main processes required for video streaming. Additional processes (e.g., for video content protection and analysis of video access rates) are discussed in the later parts of the paper (Sections 3.6–3.8, respectively).

The first step in video streaming is video content production. A raw video content generated by cameras can consume enormous storage space, which is impossible to be transmitted via current Internet speed. For instance, one second of raw video with 4K resolution occupies approximately 1 GB storage. Therefore, the generated video, firstly, has to be compressed, which is also known as *video encoding*. The concept of the video is just continuously showing a large number of frames at a certain rate

---

[i] https://livestream.com/.

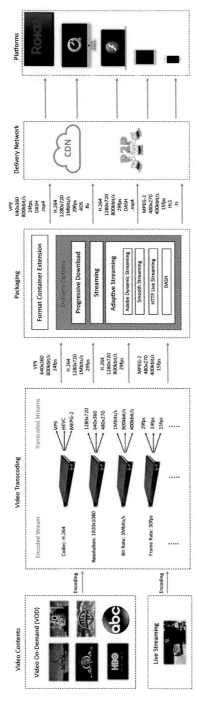

Fig. 2 A bird's-eye view to the workflow of processes performed on a video stream from the source to the viewer's device. The workflow includes compression, transcoding, packaging, and delivery processes.

(aka frame rate) to create a moving delusion. This large number of frames usually contains spatial and temporal redundancies both inside a frame and between successive frames. In the video compression process, these redundancies are removed using a certain compression standard, e.g., H.264/MPEG-4 AVC [20], VP9 [21], and H.265/HEVC [22]. The compression process encodes the raw video based on a specific compressing standard (known as *codec*), resolution, bit rate, and frame rate at a significantly smaller size.

A viewer's display device generally can decode videos with a certain compression format. Therefore, to support heterogeneous display devices, a video encoded in a certain format has to be converted (also called *transcoded*) to various formats. In the transcoding process, a video is first decoded and then encoded to another compression format. Thus, transcoding is generally a computationally intensive process. In Section 3.3, we elaborate further on the video transcoding process.

For streaming of a transcoded video to a viewer's device, the video file has to be structured to facilitate transferring and time-based presentation of the video content. Thus, the transcoded video file must be *packaged* based on a certain structure that is supported by the viewer's player. The packaging process is the basis for what is commonly known as the video file format (also, called as *container format*). The container format introduces a header that includes information about the supported streaming protocols and the rules to send video segments. ISO Base Media File Format (ISOBMFF) [23], which is the basis for *MP4* format, and 3GPP TS [24], which is the basis for *ts* format, are widely used for the packaging process.

To deliver the packaged (i.e., formatted) video files over the network, they need to use an application layer network protocols (e.g., HTTP [25], RTMP [26], and RTSP [27]). There are various *delivery techniques* that dictates the way a video stream is received and played. Progressive download [28], HTTP Live Streaming (HLS) [29], and Dynamic Adaptive Streaming over HTTP (DASH) [30] are examples of delivery techniques. Each delivery technique is based on a particular network application layer protocol. For instance, HLS and DASH work based on HTTP and Adobe Flash streaming works based on RTMP. Further details about video packaging are discussed in Section 3.4.

Once the video is packaged, it is delivered to viewers around the world through a distribution network. However, due to the Internet transmission delay, viewers located far from video servers and repositories suffer from a long delay to begin streaming [31]. To reduce this delay, content delivery networks (CDN) [32] are used to cache the frequently watched video contents at geographical locations close to viewers. Another option to reduce the startup delay is to use peer-to-peer (i.e., serverless) approaches in which each receiver (e.g., viewer's computer) acts as a server and sends video contents to another peer viewer [33]. More details of the delivery network will be discussed in Section 3.5

3. Video streaming: Challenges and solutions

3.1 Overview

In the previous section, we described the high-level workflow of processes for video streaming. However, there are various approaches and challenges to accomplish those processes. In this section, we elaborate on all issues, challenges and current solutions that are related to video streaming.

Fig. 3 presents a taxonomy of all the issues that a video streaming system needs to deal with. The taxonomy shows different types of video streaming (i.e., VOD and live streaming). It discusses variations of video transcoding and packaging processes along with important streaming protocols. The taxonomy shows possible ways of distributing video content, such as CDN [34] and P2P [35]. Video content protection is further detailed on video privacy for video streaming [36] and copyright issues, known as digital rights management (DRM) [37]. The taxonomy covers analytical studies that have been conducted to understand the viewers' behaviors and discovering their access patterns to videos streams. Last, but not the least, the taxonomy covers the storage issue for video streaming and possible strategies for video streaming repository management, such as in-house storage vs outsourcing solutions via cloud storage services.

3.2 Video streaming types

The taxonomy shown in Fig. 4 expresses possible ways a video streaming service can be provided to viewers. More specifically, video streaming service can be offered in three main fashions: *video on demand* (VOD) streaming, *live* streaming, and *live-to-VOD* streaming.

3.2.1 VOD streaming

In VOD streaming, which is also known as Over-The-Top (OTT) streaming, the video contents (i.e., video files) are already available in a video streaming repository and are streamed to viewers upon their requests. By far, VOD is the most common type of video streaming and is offered by major video streaming providers, such as YouTube, Netflix, and Amazon Prime Video.

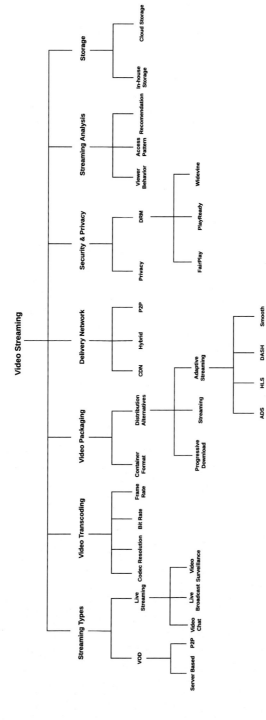

Fig. 3 Taxonomy of all aspects we need to deal with in video streaming.

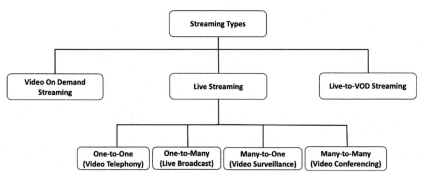

Fig. 4 Taxonomy of different types of video streaming.

Some VOD providers (e.g., Netflix and Hulu) offer professionally made videos and movies that are subscription-based and viewers are generally required to pay a monthly fee to access their service. Alternatively, other VOD services (e.g., YouTube) operate based on user-provided videos. Such services are generally advertisement-based and free of charge. VODs have also applications in e-learning systems [3], Internet television [38], and in-flight entertainment systems [39].

3.2.2 Live streaming

In live video streaming, the video contents are streamed to the viewer(s), as they are captured by a camera. Live video streaming has numerous applications, such as event coverage and video calls. The live video streaming used in different applications have minor differences that are mainly attributed to the buffer size on the sender and receiver ends [40]. In general, a larger buffer size causes a more stable streaming experience but imposes more delay. This delay can be tolerated in live broadcasting applications, however, delay-sensitive applications (e.g., video telephony) cannot bear the delay, thus need a shorter buffer size. Accordingly, live streaming can have four variations as follows:

(A) *One-to-one* (unicast) streaming is when a user streams video contents to another user. This type is primarily used in video chat and video call applications that require two live streams, one from each participant. This streaming type requires short delays to enable smooth conversation between participants. As such, these applications generally operate with a short buffer size and low picture quality to make the delay as small as possible [40]. Skype[j] and video telephony applications [41] like FaceTime[k] are instances of this type of live streaming.

(B) *One-to-many* (multicast) streaming is when one source streams video to many viewers. A well-known example of this type is live broadcasting which is currently offered by many social network applications, such as Facebook and Instagram. Jo et al. [42] conducted one of the first studies on live streaming. They identified and addressed several challenges in multicast streaming regarding signaling protocols, network stability, and viewer variations.

(C) *Many-to-one* occurs when several cameras capture scenes and send them to one viewer. The most important application for this type of streaming is multicamera video surveillance which is used for situational awareness for security purposes or natural disaster management [43]. In this type of streaming, the video contents are collected from multiple cameras and displayed on special multiscreen monitoring devices [44].

(D) *Many-to-many* streaming occurs when a group of users in different geographical locations holds a video conference. In this case, all users stream live to all others. For this streaming type, multipoint control unit (MCU) [40] method can be used to combine individual participants into a single video stream and broadcast it. Most of video chat applications, e.g., Skype and Google Hangouts, support many-to-many live streaming, in addition to one-to-many streaming.

3.2.3 Live-to-VOD streaming

In addition to live and VOD streaming, we can also consider a combination of live and VOD streaming, known as Live-to-VOD [45], as another type of streaming. In this type of streaming, which is mostly used on one-to-many live streaming, the live video stream is recorded and can be readily used in form of VOD streaming.

Using live-to-VOD streaming, viewers who are not online during the live stream can watch the video at a later time. In addition, live-to-VOD can provide VOD-like services to live stream viewers. For instance, live stream viewers can have the rewind ability. Another application of live-to-VOD is to play live video contents in different time zones. For example, using live-to-VOD, the same TV program that is live streamed at 8:00 am in the Eastern Time Zone, can be played at 8:00 am in the Pacific Time Zone.

---

[j] https://www.skype.com.

[k] https://support.apple.com/en-us/HT204380.

3.2.4 Differences in processing live and VOD streaming

Although the workflow and processes that are applied to live video streams are the same as those for VOD, there are certain differences between them. Specifically, live and VOD streaming are different on the way they are processed [12].

First, in live streaming, the video segments are processed as they are generated. This has two implications:

- The video segments have to be processed (e.g., transcoded) on-the-spot, whereas in VOD, it is possible to preprocess videos (i.e., in an off-line manner).
- There is no historic execution (i.e., processing) time information for live video segments [12]. In contrast, in VOD, each video segment is processed multiple times and the historic execution time information are available. The historic information are particularly important for the efficient scheduling of video streaming tasks.

The second difference between live and VOD streaming is the way they are treated for processing. In both live and VOD streaming, to ensure QoE, each video streaming task is assigned a deadline that must be respected. That is, the video task processing must be completed before the assigned deadline. The deadline for each video task is determined based on the presentation time of the pertinent video segment. If a task cannot meet its assigned deadline for any reason, then, in VOD, the task has to wait until it is processed [46]. In contrast, in live streaming, if a task misses its deadline, the task must be dropped (i.e., discarded) to keep up with the live streaming [12]. In other words, there is no reason to process a task whose time has passed.

The third difference between VOD and live streaming is again related to the way deadlines are assigned to video streaming tasks. In fact, in video streaming, if a task misses its deadline, all the tasks behind that, i.e., those process video segments later in the stream, should update their streaming deadlines (presentation times) accordingly. This is also known as the dynamic deadline, however, this is not the case in live streaming and the tasks' deadlines in live streaming cannot be changed.

3.3 Video transcoding

Video contents are originally encoded with one specific spatial resolution, bit rate, frame rate, and compression standard (codec). In order to play the videos on different devices, streaming service providers usually have to adjust the original videos in terms of the viewer's device and network bandwidth. The process of this adjustment is called *video transcoding* [47, 48].

Video transcoding is a compute-intensive task and needs powerful computers to process it [49]. Therefore, video transcoding is generally carried out in an off-line manner, called pretranscoding [7]. In Sections 3.3.1–3.3.4, we will elaborate different transcoding operations, respectively.

3.3.1 Bit rate

Video bit rate is the number of bits used to encode a unit time of video. Bit rate directly impacts on video quality, as higher bit rate produces better quality, while higher bit rate consumes larger network bandwidth and storage. In order to stream videos to viewers with different network conditions, streaming service providers usually convert video with multiple bit rates [50].

3.3.2 Spatial resolution

Resolution represents the dimensional size of a video, which indicates the number of pixels on each video frame. Therefore, higher resolution contains more pixels and details, as results in larger size. Video resolution usually needs to match with the screen size. Low resolution video plays on large screen will causes blurry after upsample, while high resolution plays on small screen is just a waste of bandwidth since viewer usually will not notice the difference due to the limited pixels on the screen. To adapt to the diverse screen size on the market, original videos have to be transcoded to multiple resolutions [51].

3.3.3 Frame rate

When still video frames plays at a certain speed rate, human visual system will feel the object is moving. Frame rate indicates number of video frames shown per second. Videos or films are usually recorded at high frame rate to produce smooth movement, while devices may not support such high frame rate. Therefore, in some cases, videos have to be reduced to a lower frame by removing some frames. On the other hand, increasing frame rate is more complicated than reducing frame rate, since it have add nonexistent frames. Overall, to be adaptive to larger scale device, video are transcoded to different frame rates [52].

3.3.4 Video compression standard

Video compression standard is the key to compress a raw video, the encoding process mainly goes through four steps: prediction, transformation, quantization, and entropy coding, while decoding is a just reverted encoding process. With different codecs manufactured on different devices (e.g., DVD player with MPEG-2 [53], BluRay player with H.264 [54], 4K TV with HEVC [22]), an encoded video may have to be converted to the supported codec on that device. Changing codec is the most compute-intensive type of transcoding [49] since it has to decode the bitstream with the old codec first and then encode it again with the new codec.

3.4 Video packaging

Transmitting an encoded/transcoded video file from server to viewer involves multiple layers of network protocols, namely physical layer, data link layer, network layer, transport layer, session layer, presentation layer, and application layer [55]. The protocols of these layers dictate video packaging details, such as stream file header syntax, payload data, authorization, and error handling. Since video streaming protocols operate under the application layer, they potentially can use different protocols in the underlying layers to transmit data packets.

Choosing the right streaming technology requires understanding pros and cons of the streaming protocols and video packaging (aka container formats). In this section, we discuss three popular streaming technologies plus the streaming protocols and container formats required for each one of the technologies.

3.4.1 Progressive download

Back in old days, when online video streaming was not practical, a video could not be viewed until it was completely downloaded on the device. This implies that viewers usually had to wait for a considerable amount of time (from minutes to even hours) before watching the video. Progressive download resolved this issue by allowing a video to be played as soon as the player's initial buffer is filled by segments of the video. This reduces the waiting time down to 3–10 s to begin watching a video [56].

Due to the downloading feature, progressive download can face three problems. First, since a video is downloaded linearly, if the viewer's network bandwidth is too low, the viewer cannot move forward a video until that part is fully downloaded. Second downside of progressive download is that if a video file is fully downloaded, but viewer stops watching in the middle, the rest bandwidths are wasted. Third, copyright is problematic in progressive download because the whole video is downloaded on the viewer's storage device. Progressive download utilizes HTTP protocol that itself operates based on the TCP protocol, which provides better reliability and error-resilience than UDP, but it incurs a high network latency [57]. These inherent drawbacks of HTTP-based progressive download raised the need to a dedicated technology for video streaming.

3.4.2 Dedicated protocol for video streaming

To avoid problems of progressive download, a dedicated protocol for real-time streaming (known as RTP) [58] was created. This protocol delivers video contents from a separated streaming server. While traditional HTTP servers handle web requests, streaming servers only handle streaming. The connection is initiated between the player and the streaming server whenever a viewer clicks on a video in a web page. The connection persists until the video terminates or the viewer stops watching it. In comparison to stateless HTTP, RTP protocol is considered stateful because of this persistent connection.

Because of the persistent connection, dedicated streaming protocols allow random access (e.g., fast forward) within the streamed video. In addition, they allow adaptive streaming, in which multiple encoded video streams could be delivered to different players based upon available bandwidth and processing characteristics. The streaming server can monitor the outbound flow, so if the viewer stops watching the video, it stops sending video packet to the viewer.

Video content in the streaming server is split into small chunks, whenever these chunks are sent to the viewers, they are cached at the local device and will be removed after they are played. This feature offers freedom to viewer to move back and forth within the streamed video. It also protects the copyright of the video content. Although streaming technology was attractive in the beginning, its drawback appeared after deployment. A streaming protocol, e.g., RTMP [26] used by Adobe Flash, utilizes different port numbers from HTTP. As such, RTMP packets can be blocked with some firewalls. The persistent connection between the streaming server and viewer players increases the network usage and causes limited scalability for streaming servers.

3.4.3 Adaptive streaming

To address the limitations of previous streaming technologies, HTTP-based streaming solutions came back to the forefront of streaming technology-adaptive streaming. All adaptive streams are broken into chunks and stream separate videos. There is no persistent connection between the server and the player. Instead of retrieving a single large video file in one request, adaptive streaming technology retrieves a sequence of short video files in an on-demand basis.

Adaptive streaming has the following benefits: First, like streaming server, there is no network wastage, because the video content is delivered on the go. Therefore, one HTTP server can efficiently serves several streams. Second, HTTP-based streaming is delivered through HTTP protocol, which avoids the firewall issue faced by RTMP. Third, it costs less than using streaming server. Fourth, it can scale quickly and effectively to serve more viewers. Fifth, seeking inside the stream is no more an issue. When the viewer moves the player forwarder, the player just retrieves the exact video segments as opposed to the entire video up to the requested point.

There are four main adaptive streaming protocols, namely MPEG Dynamic Adaptive Streaming over HTTP (DASH) [30], Apple HTTP Live Streaming (HLS) [59], and Microsoft Smooth Streaming [60].

MPEG-DASH delivers ISOBMFF [61] video segments. It defines a media presentation description (MPD) XML document to provide the locations of video streams, so that players know where to download them. The media segments for DASH is delivered with formats either based on the ISOBMFF [23] or Common Media Application Format (CMAF) standards [62].

Apple's HLS is well known and implemented on the Apple devices (and nowadays on Android devices too). It utilizes a M3U8 master manifest to include multiple media playlists, each playlist represents one stream version and it contains the location of all the video segments for this stream. HLS video segment uses either MPEG-2 Transport Stream (M2TS) [63] or CMAF [62] for H.264 encoded videos, and ISOBMFF for HEVC encoded videos.

Smooth Streaming has two separated manifest files, namely SMIL server manifest file and client manifest file. They are all defined in XML format documents. Smooth streaming also delivers video segment with a format (known as ISMV) based on ISOBMFF.

These four adaptive streaming technologies have empowered streaming service providers to deliver the video contents to viewers smoothly even under low bandwidth Internet connection. However, to support all viewers' platforms, stream providers have to deploy and maintain all these four streaming protocols that subsequently increases complexity and costs. The supported platforms of these four protocols are shown in Table 1

Table 1 Adaptive streaming supported platforms.

	Desktop player	Mobile device support	OTT support
MPEG-DASH	dash.js, dash.as, GPAC	Windows, Android Phone	Google TV, Roku, Xbox 360
Apple HTTP Live Streaming (HLS)	iOS, Mac OSX, Flash	iOS/Android3.0+	Apple TV, Boxee, Google TV, Roku
Smooth Streaming	Silverlight	Windows Phone	Google TV, Roku, Xbox 360

3.5 Video streaming delivery networks

3.5.1 Content delivery networks

The goal of CDN technology is to reduce the network latency of accessing web contents. CDNs replicate the contents to geographically distributed servers that are close to viewers [64, 65]. Considering that large size of video contents usually takes a long transmission time, using CDNs to cache video contents close to viewers reduces the latency dramatically. Netflix, as one of the largest stream providers use three different CDN providers (Akamai, LimeLight, and Level-3) to cover all viewers in different regions [66]. The transcoded and packaged video contents are replicated in all three CDNs.

Streaming video contents through CDN has been studied in earlier works [67–69]. Cranor et al. [68] proposed an architecture (called PRISM) for distributing, storing, and delivering high-quality streaming content over IP networks. They proposed a framework to support VOD streaming content via distribution networks services (i.e., CDNs). Wee et al. [69] presented an architecture for mobile streaming CDN which was designed to fit the mobility and scaling requirements. Apostolopoulos et al. [67] proposed multiple paths between nearby edge servers and clients in order to reduce latency and deliver high-quality streaming. Benkacem et al. [70] provided an architecture to offer CDN slices through multiple administrative cloud domains. Al-Abbasi et al. [71] proposed a model for video streaming systems, typically composed of a centralized origin server, several CDN sites, and edge-caches located closer to the end users. Their proposed approach focused on minimizing a performance metric, stall duration tail probability (SDTP), and present a novel and efficient algorithm accounting for the multiple design flexibilities. The authors demonstrated that the proposed algorithms can significantly improve the SDTP metric, compared to the baseline strategies.

Live streaming CDN is discussed in [72–74] to improve scalability, latency quality, and reliability of the service.

3.5.2 Peer to peer (P2P) networks

P2P networks enable direct sharing of computing resources (e.g., CPU cycles, storage, and content) among peer nodes in a network [33]. P2P networks are designed for both clients and servers to act as peers. They can download data from same nodes and upload them to other nodes in the network. P2P networks can disseminate data files among users within a short period of time. P2P networks are extensively utilized for video streaming services as well.

P2P streaming is categorized into two types, namely tree-based and mesh-based. Tree-based P2P structure distributes video streams by sending data from a peer to its children peers. In mesh-based P2P structure, the peers do not follow a specific topology, instead, they function based on the content and bandwidth availability on peers [35]. One drawback of using tree-based P2P streaming is the vulnerability to peers' churn. The drawback of deploying mesh-based P2P streaming structure is video playback quality degradation, ranging from low video bit rates and long startup delays, to frequent playback freezes. Golchi et al. [75] proposed a method for streaming video in P2P networks. Their approach uses the algorithm of improved particle swarm optimization(IPSO) to determine the appropriate way for transmitting data that ensures service quality parameters. They showed that the proposed approach works efficiently more than the other related methods.

Fig. 5 shows two popular variations of tree-based P2P structure, namely single tree-based and multitree based structures. In *single tree-based streaming*, shown in Fig. 5A, video streaming is carried out at the application layer and the streaming tree is formed by participating users. Each viewer joins the tree at a certain level and receives the video content from its parent peer at the above level. Then, it forwards the received video to its children peers at the below level [76, 77].

In *multitree streaming* (aka mesh-tree), shown in Fig. 5B, a peer develops the peering connection with other close peers. Then, the peer can potentially download/upload video contents from/to multiple close peers simultaneously. If a peer loses the connection to the network, the receiving peer could still download video from remaining close peers. Meanwhile, the peer can locate new peers to maintain a required level of connectivity. The strong peering in mesh-based streaming systems supports them to be robust against peer churns [78, 79]. Ahmed et al. [80] proposed a multilevel multioverlay

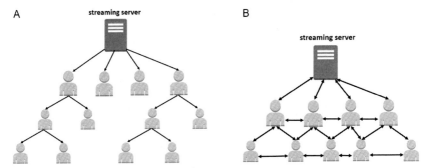

Fig. 5 Ways to achieve peer-to-peer (P2P) video streaming (A) single tree-based streaming; (B) mutlitree streaming.

hybrid peer-to-peer live video system that offers to the Online Games players the way of streaming the video simultaneously and enables to watch the game videos of other players. Their approach aimed to reduce the transmission rate while increasing the number of peers and the delivery and reliability of data are guaranteed on time.

Tree-based P2P VOD streaming shares the video stream on the network and achieves fast streaming. The video stream is divided into small size data block. The server disperses the data blocks to different nodes. The nodes download their missing blocks from their neighboring peers [81].

Guo et al. [82] proposed an architecture that uses the P2P approach to cooperatively stream video by only relying on unicast connections among peers. Xu et al. [83] proposed an approach based on binary tree strategy to distribute VOD P2P networks. Their approach divides the videos into segments to be fetched from several peers. Yiu et al. [84] proposed VMesh distributed P2P VOD streaming approach. In their approach, videos are split into segments and then stored at the storage of the peers locally. Their proposed design presents to peers an ability to forward, backward, pause, and restart during the playback. Cheng et al. [85] proposed a topology enhancing video cassette recording (VCR) functions for VOD services in the networks. Their approach allows a peer to achieve fast seeks relocation by keeping close neighbors and remote them in a set of rings. Xu et al. [86] proposed a scheme based tree to make user interacting in the P2P streaming. The proposed scheme presents an advantage to support the user's requests asynchronously while maintaining high resilience.

The main advantages of P2P are low cost and flexibility for scalability, however, it suffers from instability in QoS. Therefore, researchers proposed to combine the advantages of both CDN and P2P in one system.

Accordingly, several hybrid systems were developed to combine P2P and CDN for content streaming. Afergan et al. [87] proposed an approach which utilizes CDN to build CDN-P2P streaming. In their proposed design, they optimized dynamically the number and locations of replicas for P2P service. Xu et al. [88] presented a scheme that is formed by both CDN and P2P streaming. They showed the efficiency of their approach by reducing the cost of CDN without impacting the quality of the delivered videos.

3.6 Video streaming security

3.6.1 Privacy

With the ubiquity of video streaming on a wide range of display devices, the privacy of the video contents has become a major concern. In particular, live contents either in form of video surveillance or user-based live streaming capture places and record many unwanted/unrelated contents. For instance, a person who live streams from a street, unintentionally may capture plate number of vehicles passing that location. Therefore, video streaming systems can compromise the privacy of people (e.g., faces and vehicles tags). Various techniques have been developed to protect the privacy of live video contents.

Dufaux et al. [89] introduced two techniques to obscure the regions of interests while video surveillance systems are running. Zhang et al. [36] came up with a framework to store the privacy information of a video surveillance system in form of a watermark. The proposed model embeds a signature into the header of the video. Moreover, it embeds the authorized personal information into the video that can be retrieved only with a secret key. Another research, conducted by Carrillo et al. [90], introduces a compression algorithm for video surveillance system, which is based on the encryption concept. The proposed algorithm protects privacy by hiding the identity revealing features of objects and human. Such objects and human identity could be decrypted with decryption keys when an investigation is requested.

Live video contents commonly are transmitted via wireless media which can be easily intercepted and altered. Alternatively, DOS attacks can be launched on the live video traffics [91]. As such, in [92, 93] algorithms are provided to distinguish between packets that were damaged because of noise or an attack. In the former case, the errors must be fixed while in the latter the packet must be resent. The algorithm counts the number of 1s or 0s in a packet before the packet is send and uses that count to generate a message authentication code (MAC). The MAC is appended to the end of the packet and sent over the network. When the packet is received, the MAC is calculated again, and the two codes are compared. If the differences in MACs is past a certain threshold, the past is marked as malicious and discarded. Because the MAC is also sent over the network, the algorithm will detect bit errors in both the packet and the MAC.

3.6.2 Digital rights management

Another security aspect in video streaming is the copyright issue. This is particularly prominent for subscription-based video streaming services (such as Netflix). DRM is the practice of securing digital contents, including video streams, to prevent unlawful copying and distribution of the protected material. Specifically, its goals are typically centered around controlling access to the content, enforcing rules about usage, and identifying and authenticating the content source. As such, contents protected with DRM are typically sold as a license to use the contents, rather than the content itself. DRM solutions to meet these goals often incorporate tools such as encryption, watermarking, and cryptographic hashing [94].

The process of DRM starts with encryption, video contents are encrypted and stored in the video repository and DRM providers keep the secret keys. Fig. 6 summarizes the steps to stream a video protected with

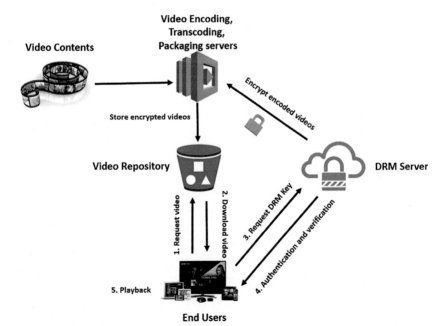

Fig. 6 Workflow of Digital Right Management (DRM) to support of video streaming security.

DRM. Upon request to stream an encrypted video, after downloading it, the secret key is requested from the DRM provider server to decrypt the video.

Currently, there are three major DRM technologies, namely Apple's Faireplay [95], Microsoft's PlayReady [96], and Google's Widevine [96]. These technologies are able to serve most of the devices and platforms in the market. To cover all devices, streaming service providers have to encrypt every video content with three different DRM technologies and store them in the repository. The three DRM technologies supported platforms are shown in Table 2.

With the increasing demand to stream DRM-protected VOD, an off-line application-driven DRM interceptor has been proposed in [97] that is the awareness of the network connection status of the client device. In this case, if the interceptor application decides that the client device is offline, it requests the license/key for the protected content. The license/key is controlled by the interceptor application. Accordingly, the requests of license/key are handled by the interceptor application that retrieves them from a locally data-store, and then send the key/license to the DRM module.

3.7 Analysis of video streaming statistics

3.7.1 Impact of quality on viewers' behavior

Previous studies show that the quality of video streaming impacts on viewers' reception to the video and the revenue of the stream provider [31]. According to the studies, starting video streaming with delay and interruption during video streaming significantly increases the possibility of abandoning watching the videos.

Florin et al. [98] addressed the impact of video quality on viewers' interest. They claimed that the percentage of time spent on buffering (i.e., buffering ratio) has the largest impact on the user interest across all content types. The average bit rate of live streaming has a significant impact on user abandonment of watching the video. Video streaming providers should attempt to maximize viewer engagement by minimizing the buffering time and rate and increasing the average bit rate.

3.7.2 Access pattern to video streams

Analysis of access pattern to videos shows that the access pattern of videos streams in a repository follows a long-tail distribution [10]. That is, few popular videos, known as *hot* videos that construct around 5% of the repository, are accessed very often while a large portion of nonpopular videos are accessed infrequently.

Table 2 DRM supported platforms.

	Chrome	FireFox	IE 11	Safari	Android	iOS	Windows phone	ChromeCast	Roku	Apple TV	Xbox
Fairplay				✓		✓				✓	
PlayReady			✓	✓			✓	✓	✓		✓
Widevine	✓	✓			✓			✓			

The studies also reveal that viewers are typically interested in recently posted videos. Moreover, for new videos, the popularity fluctuates significantly while the popularity of old videos does not fluctuate significantly [99].

Video access rate indicates how many times a video is accessed by a user, however, it does not implicate if the accessed video stream is played to end of it or not. In fact, recent studies (e.g., [11]) showed that, the beginning segments of a video stream are played more frequently than the rest of it. Miranda et al. [11] revealed that in a video stream, the views are distributed following a long-tail distribution. More specifically, the distribution of views of the segments (GOPs) in a video stream can be calculated by the Power-law [100] model.

The access rate of GOPs in all video streams in a repository does not necessarily follow long-tail pattern as stated earlier. There are video streams whose some GOPs in the middle or end of the video stream are accessed more frequently than other GOPs. An example of a soccer match streaming can show GOPs with a tremendously higher access rate where a player scores a goal. We define this type of video streams as those with nonlong-tail access pattern [101].

3.7.3 Video streaming-based recommendation

As mentioned above, provided the access pattern of video streams for a given viewer, stream providers are able to predict the video categories the viewer prefers and recommend them at the front page. The same strategy can be used for video recommendation to the viewers located in the same geographical area. To improve the accuracy of prediction and recommendation, machine learning [102] and deep learning [103] approaches have widely been applied for this purpose. The most successful example is Netflix and YouTube machine-learning-based recommendation systems as explained in [104].

3.8 Storage of video repositories

Video streaming repositories are growing in size, due to the increasing number of content creation sources, diversity of display devices, and the high qualities viewer's desire. This rapid increase in the size of repositories implies several challenges for multimedia storage systems. In particular, video streaming storage challenges are threefold: capacity, throughput, and fault tolerance.

One of the main reasons to have a challenge in video streaming storage capacity is the diversity of viewers' devices. To cover increasingly diverse display devices, multiple (more than 90) versions of a single video should be generated and stored. However, storing several formats of the same video implies a large-scale storage system. Previous studies provide techniques to overcome storage issues of video streaming. Miao et al. [105] proposed techniques to store some frames of a video on the proxy cache. Their proposed technique reduces the network bandwidth costs and increases the robustness of streaming video on poor network conditions.

When a video is stored on a disk, concurrent accesses to that video are limited by the throughput of the disk. This restricts the number of simultaneous viewers for a video. To address the throughput issue, several research studies have been undertaken. Provided that storing video streams on a single disk has a low throughput, multiple disk storage are configured to increase the throughput. Shenoy et al. [106] propose data stripping where a video is segmented and saved across multiple storage disks to increase the storage throughput. Videos streams are segmented into blocks before they are stored. The blocks can be stored one after another (i.e., contiguously) or scattered over several storage disks. Although contiguous storage method is simple to implement, it suffers from the fragmentation problem. The scattered method, however, eliminates the fragmentation problem with the cost of more complicated implementation.

Scattering video streams across multiple disks and implementing data striping and data interleaving methods improves reliability and fault tolerance of video streaming storage systems [107].

4. Cloud-based video streaming

4.1 Overview

In this section, we discuss how offered cloud services can be useful for different processes in video streaming. Broadly speaking, we discuss computational services, networking services, and storage services offered by cloud providers and are employed by video stream providers. We also elaborate on possible options within each one of the cloud services.

4.2 Computing platforms for cloud-based video streaming

Upon arrival of a streaming request, the requested video stream is fetched from the cloud storage servers and the workflow of actions explained in Section 2.2 are performed on them, before streaming them to the viewers. These processes are commonly implemented in form of independent services, known as microservices [108], and are generally deployed in a loosely coupled manner on independent servers in cloud datacenters. Services like web server, video ingestion, encoding, transcoding, and packaging are examples of microservices deployed in datacenters for video streaming. For the sake of reliability and fault tolerance, each of these microservices is deployed on multiple servers and form a large distributed system. A load balancer is deployed for the servers of each microservice. The load balancer assigns streaming tasks to an appropriate server with the goal of minimizing latency and to cover possible faults in the servers.

The aforementioned microservices are commonly implemented via container technologies [109], possibly using serverless computing paradigm [110]. Docker containers scale up and down much faster than VMs and have faster startup (boot up) times that gives them an advantage to VMs in handling fluctuating video streaming demands. In addition, Docker containers are used in video packaging, handling arriving streaming requests (known as request ingestion), and inserting advertisements within/between video streams.

Sprocket [110] is a serverless system implemented based on AWS Lambda [111] and enables developers to program a series of operations over video content in a modular and extensible manner. Programmers implement custom operations, ranging from simple video transformations to more complex computer vision tasks, and construct custom video processing pipelines using a pipeline specification language. Sprocket then manages the underlying access, encoding and decoding, and processing of video and image content across operations in a parallel manner. Another advantage of deploying serverless (aka function-as-a-service) computing paradigm, such as AWS Lambda, for video streaming is to relieve stream providers from scheduling, load balancing, resource provisioning, and resource monitoring challenges.

Apart from the type of computing platform (e.g., VM-based and container-based) for video streaming, the type of machines provisioned to process streaming tasks are also influential in the latency and incurred cost of streaming. For instance, clouds offer various VM types with diverse configurations (e.g., GPU base, CPU base, IO base, and Memory base), or various reservation types (e.g., on-demand, spot, and advance reservation).

4.3 Cloud-based video transcoding

To cover viewers with heterogeneous display devices that work with diverse codecs, resolutions, bit rates, and frame rates, video contents usually have to be transcoded to several formats. Video transcoding process takes major computing power and time in the whole video streaming workflow [49]. The Methods and challenges for video transcoding have been studied by Vetro et al. [48] and Ahmad et al. [47]. In the past, streaming service providers had to maintain large computing systems (i.e., in-house datacenters) to achieve video transcoding. However, due to the update and maintenance costs of in-house datacenters, many streaming service providers have chosen to outsource their transcoding operations to cloud servers [7, 9, 12, 49, 112]. Extensive computational demand of video transcoding can potentially impose a significant cost to stream providers. As such, it is important that stream providers apply proactive methods to transcode videos in their repositories.

A taxonomy of various cloud-based video transcoding is shown in Fig. 7. Cloud-based solutions for transcoding VODs are either based on creating several versions of the original video in advance (aka pretranscoding [16–18, 113, 114]) or transcoding videos upon viewer's request (aka on-demand transcoding [9, 46, 115]). In considering the long-tail access pattern to video streams on the Internet has made the on-demand approach an attractive option for stream providers. However, pure on-demand transcoding approach can increase latency and even the incurred cost for stream providers [112]. Therefore, approaches have been introduced to perform pretranscoding in a more granular level. That is, pretranscoding only parts of a video streams (e.g., few important GOPs). This approach is known as partial pretranscoding [101].

According to Darwich et al. [101], partial pretranscoding can be carried out either in a deterministic or nondeterministic manner. Considering the fact that the beginning of video streams are generally watched more often [11], in the deterministic approach, a number of GOPs from the beginning

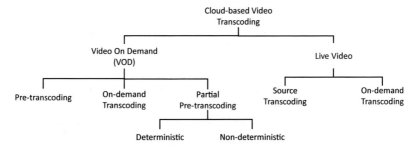

Fig. 7 Approaches to perform video transcoding using cloud services.

of the video are pretranscoded, and transcode the rest in an on–demand manner [112]. Alternatively, in the nondeterministic approach [101], pre-transcoding is not limited to the beginning of video streams and can be performed on any popular GOP, disregarding its position in the stream. Although the nondeterministic approach is proven to be more efficient, it imposes the overhead of maintaining and processing view metadata for each GOP.

Procuring various video formats in live streaming can be achieved by camera, upon video production (known as source transcoding). However, due to the high latency and inefficiency of source transcoding, live videos are commonly transcoded in the run time [12, 116, 117].

Research have been focused on video segmentation [16, 113], load balancing [17, 114], and resource provisioning [18, 114], with the goal of maximizing throughput of the pretranscoding operation. Alternatively, in on–demand transcoding, the goal is to respect QoE of viewers in form of minimizing the response time of the transcoding operation. More specifi-cally, each segment (GOP) of the video needs to be transcoded before its presentation time (aka deadline).

To use cloud services efficiently, it is crucial to realize the performance difference among video transcoding operations on different VM types. Li et al. [49] have evaluated the performance of various transcoding operations on heterogeneous VMs to investigate the key factors on transcoding processing time. They identified the affinity (i.e., match) of various VM types (e.g., GPU vs CPU and Memory-based) with different transcoding operations. They show that GOP frame numbers is the most influential fac-tor on transcoding time. In fact, frame numbers in a GOP can imply its con-tent type. Videos with fast-motion content include numerous GOPs with few frames and vice versa. The execution time of each small GOP is usually low, and therefore it can be transcoded on cost-efficient VM types without compromising performance. In addition, Li et al. provide a suitability model of heterogeneous VM types for various transcoding operations in consider-ation of both performance and cost factors. Transcoding time estimation is needed to improve the quality of scheduling and VM provisioning on the cloud. Deneke et al. [15] propose a machine learning model to predict the transcoding time based on the video characteristics (e.g., resolution, frame rate, and bit rate).

To further cut the cost of utilizing cloud services for on-demand video transcoding, Li et al. [7, 9] propose cloud-based video streaming engine (CVSE) that operates at the video repository level and pretranscodes only hot videos while retranscodes rarely accessed videos upon request. Li et al. define desired QoE of streaming as minimizing video streaming startup delay via prioritizing the beginning part of each video. The engine is able to provide low startup delay and playback jitter with proper scheduling and resource provisioning policy.

Transcoding video in an on-demand way does reduce the storage cost, however, it incurs computing cost. How to balance these two operations to minimize the incurred cost is another challenge. Jokhio et al. [118] present a trade-off method to balance the computation and storage cost for cloud-based video transcoding. The method is mainly based on the time and frequency for a given video to be stored and retranscoded, respectively. Compared to Jokhio's work, Zhao et al. [119] also take the popularity of the video into consideration. Darwich et al. [101] propose a method to partially pretranscode video streams depending on their degree of hotness. For that purpose, they define a method based on the hotness of each GOP within the video.

Barais et al. [120] propose a microservice architecture to transcode videos at a lower cost. They treat each module (e.g., splitting, scheduling, transcoding, and merging) of transcoding as a separate service, and running them on different dockers. To reduce the computational cost of on-demand transcoding, Denninnart et al. [121] propose to aggregate identical or similar streaming microservices. Identical streaming microservices appear when two or more viewers stream the same video with the same configurations. Alternatively, similar streaming microservices appear when viewers stream the same video but with different configurations or even different operations. An example of identical microservices can be when two or more viewers stream the same video for the same type of device (e.g., smartphone). However, when the viewers stream the same video on distinct devices (e.g., smartphone vs TV) that have different resolution characteristics, the video has to be processed to create two different resolutions, which creates microservices with various configurations. Interestingly, during peak times the method becomes more efficient because it is more likely to find similarity between microservices.

Cloud-based video transcoding has also been widely used in live streaming [12, 116]. Lai et al. [116] propose a cloud-assisted real-time transcoding mechanism based on HLS protocol, it can analyze the online quality between client and server without changing the HLS server architecture, and provides the good media quality. Timmerer et al. [122] present a live transcoding and streaming-as-a-service architecture by utilizing cloud infrastructure, it is able to take the live video streams as input and output multiple stream versions based on the MPEG-DASH [117] standard.

4.4 Cloud-based video packaging

Video packaging is computationally lighter than video transcoding, therefore, it is beneficial and more feasible to process them in an on-the-fly by utilizing cloud services (VMs or containers) [123]. For VOD streaming, video contents are usually pretranscoded to different renditions (formats), then each rendition is packaged into various versions to meet different streaming protocol requirements (e.g., DASH, HLS, Smooth).

Instead of statically packaging transcoded videos into different protocol renditions and store them in the repository (i.e., prepackaging), dynamic video packaging packages video segments based on the device's supported protocols. The whole process only takes milliseconds [124], and viewers do not generally notice it, however, it saves a significant storage cost for video streaming providers.

Due to the lightweight nature of the packaging process, container services are generally used for their implementation in cloud data centers.

4.5 Video streaming latency and cloud-based delivery networks

Clouds are known for their centrality and high latency communication to users [125]. To reduce the network latency and to reduce the load of requests from the servers, video contents are normally cached in the CDN such as Akamai. The CDNs are distributed and located geographically close to viewers.

A workflow of the whole video streaming process is shown in Fig. 8. Upon a viewer's request to stream a video, the request is first received by a web server in the cloud datacenter. If the requested video is already cached in the CDN, the web server will send back a manifest file which informs the viewer's computer about the CDN that holds the video files. Then, the viewer sends another request to the CDN and stream the video. However, if the requested video is not cached in the CDN, the web server has to process and send the content from cloud storage to the CDN. Then, the server sends a copy of the file that includes the CDN address to the viewer to start streaming from the CDN.

CDNs that are offered in form of a cloud service are known as cloud CDN. Compared to traditional CDNs, cloud CDNs are cost-effective

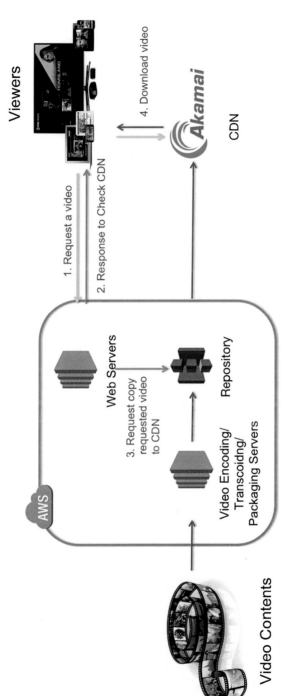

Fig. 8 Workflow of actions taken place for low latency video streaming.

and offer low latency services to content providers without having their own infrastructure. The users are generally charged based on the bandwidth consumption and storage cost [126].

Hu et al. [127] presented an approach using the cloud CDN to minimize the cost of system operation while maintaining the service latency. Based on viewing behavior prediction, Hu et al. [128] investigated the community-driven video distribution problem under cloud CDN and proposed a dynamic algorithm to trade-off between the incurred monetary cost and QoS. Their results came with less operational cost while satisfying the QoS requirement.

Jin et al. [129] proposed a scheme that offers the service of on–demand virtual content delivery for user-generated content providers to deliver their contents to viewers. The proposed approach was developed using a hybrid cloud. Their scheme offered elasticity and privacy by using virtual content delivery service with keeping the QoE to user-generated content providers.

Li et al. [130] proposed an approach for partial migration of VoD using the hybrid cloud deployment. Their proposed solution allows the requests of user to be partly served based on the self-owned servers and partly used the cloud. Their proposed migration approach (active, reactive, and smart strategies) helps the hybrid cloud to save up to 30% bandwidth expenses compared to the client/server mode. Researchers at Microsoft conducted an experiment within Microsoft public cloud CDN, Windows Azure, to demonstrate the benefits of CDN integration with the cloud. The results show a significant gain in large data download by utilizing CDN in cloud computing [131].

4.6 Cloud storage for video streaming

The rapid growth of video streaming usage in various applications, such as e-learning, video surveillance and situational awareness, and on various forms of mobile devices (e.g., smartphones, tablets, laptops) has created the problem of *big video data* [132]. The fast growth of video contents on the Internet requires massive storage facilities. However, the current storage servers face scalability and reliability issues, in addition to the high maintenance and administration cost for storage hardware. The cloud storage services provide a solution for scalability, reliability, and fault tolerance [133]. As such, major streaming service providers (e.g., Netflix and Hulu) have relied entirely on cloud storage services for their video storage demands.

Table 3 provides a comparison of different storage solutions for video streaming in terms of accessibility of viewers to the same video stream,

Table 3 Comparison of different technologies for storing video stream.

	Accessibility	Capacity	Scalability	Reliability	Cost	QoE
In house storage	Limited	Large	No	No	High	High
Cloud storage (outsourcing)	High	Large	Yes	Yes	Low	Low
Content delivery network (CDN)	High	Small	No	Yes	High	High
Peer 2 Peer	High	Large	Yes	Yes	Low	Low
Hybrid CDN-P2P	High	Small	Yes	Yes	Low	High

available capacity (storage space), scalability, reliability, incurred cost, and QoS. It is worth noting that although CDN is not a storage solution, but it can be used to reduce the storage cost by caching temporally hot video streams. Therefore, we consider it in our comparison table.

On-demand processing of video streaming is one effective method to reduce the storage volume and cost. This is particularly important when the long-tail access pattern to video streams is considered [101]. That is, except for a small portion of video streams that are hot the rest of videos are rarely accessed. Gao et al. [134] propose an approach that partially pretranscodes video contents in the Cloud. Their approach pretranscodes the beginning segments of video stream and which are more frequently accessed, while transcoding the remaining contents video stream upon request, this results to a reduction of storage cost. They demonstrated that their approach reduces 30% of the cost compared to pretranscoding the whole contents of video stream.

Darwich et al. [101] proposed a storage and cost-efficient method for cloud-based video streaming repositories based on the long-tail access patterns. They proposed both repository and video level solutions. In the video level, they consider access patterns in two cases: (A) when it follows a long-tail distribution and (B) when the video has random (i.e., nonlong-tail) access pattern. They define the cost-benefit of pretranscoding for each GOP and determine the GOPs that need to be preprocessed and the ones that should be processed in an on-demand manner.

Krishnappa et al. [135] proposed strategies to transcode the segments of a video stream requested by users. To keep a minimized startup delay of video streams when applying online strategies on video, they came up with an approach to predict the next video segment that is requested by the user.

They carried out their prediction approach by implementing Markov theory. Their proposed strategy results a significant reduction in the cost of using the cloud storage with high accuracy.

5. Summary and future research directions

5.1 Summary

Video streaming is one of the most prominent Internet-based services and is increasingly dominating the Internet traffic. Offering high quality and uninterrupted video streaming service is a complicated process and involves divergent technologies—from video streaming production technologies to techniques for playing them on a widely variety of display devices. With the emergence of cloud services over the past decade, video stream providers predominantly have become reliant on various services offered by the clouds.

In this study, *first*, we explained the workflow and all the technologies involved in streaming a video over the Internet. We provided a bird's-eye-view of how these technologies interact with each other to achieve a high QoE for viewers. *Second*, we provided details on each of those technologies and challenges the video streaming researchers and practitioners are encountering in those areas. *Third*, we reviewed the ways various cloud services can be leveraged to cope with the demands of video streaming services. We surveyed and categorized different research works undertaken in this area.

The main application of cloud services for video streaming can be summarized as: *(A)* Cloud computational services via VM, containers, or function (i.e., serverless computing) paradigms. Computational services can be used for processing video transcoding, video packaging, video encryption, and stream analytics; *(B)* Cloud network services via cloud-based CDN to reduce video streaming latency, regardless of viewers' geographical location; and *(C)* Cloud storage services to store video streaming repositories and enable persisting multiple versions of each video to support a wide range of display devices.

5.2 Future of cloud-based video streaming research

Although cloud services have been useful in resolving many technical challenges of video streaming, there are still areas that either remain intact or require further exploration by researchers and practitioners. In the rest of

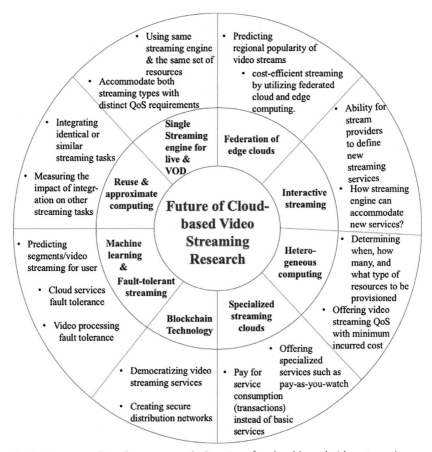

Fig. 9 Summary of the future research directions for cloud-based video streaming.

this section, we discuss several of these areas that we believe addressing them will be impactful in the future of video streaming industry. A summary of the future directions are provided in Fig. 9.

1. *Interactive video streaming using clouds.*

Interactive streaming is defined as to provide the ability for video stream providers to offer any form of processing, enabled by video stream providers, on the videos being streamed [43, 136]. For instance, a video stream provider in e-learning domain may need to offer video stream summarization service to its viewers. Another example can be providing a service that enables viewers to choose sub-titles of their own languages. Although some of these interactions are currently provided by video stream providers,[1] there is still not a generic video streaming engine that can be extended dynamically by the video provider to offer new streaming services.

Since there is a wide variety of possible interactions that can be defined on a video stream, it is not feasible to process and store video streams in advance. Instead, they have to be processed upon the viewer's request and in a real-time manner. It is also not possible to process the video streams on viewers' thin-clients (e.g., smartphones), due to energy and compute limitations [137]. Providing such a streaming engine entails answering several researches and implementation problems including:

- How to come up with a video streaming engine that is extensible and a high-level language for users to extend the engine with their desired services and without any major programming effort?
- How does the streaming engine can accommodate a new service? That implies answering how the streaming engine can learn the execution time distribution of tasks for a user-defined interaction? How can the streaming engine provision cloud resources (e.g., VMs or containers) and schedule video streaming tasks on the allocated resources so that the QoE for viewers is guaranteed and the minimum cost is imposed to the stream provider?

2. *Harnessing heterogeneity in cloud services to process video streams.*

Current cloud-based video stream providers predominantly utilize homogeneous cloud services (e.g., homogeneous VMs) for video stream processing [9, 12]. However, cloud providers own heterogeneous machines and offer heterogeneous cloud service [7]. For instance, they provide heterogeneous VM types, such as GPU, CPU-Optimized, Memory-Optimized, IO-Optimized in Amazon cloud [7]. The same is with heterogeneous storage services (e.g., SSD and HDD storage services). Heterogeneity may also refer to the reliability of the provisioned services. For instance, cloud providers offer *spot* and *reserved* compute services that cost remarkably lower than the normal compute services [138].

Traditionally, elastic resource provisioning methods for cloud determine *when* and *how many* resources to be provisioned. By considering heterogeneity, a third dimension emerges: *what type* of resources should be provisioned? The more specific questions that must be addressed are: how can we harness cloud services to offer video streaming with QoE considerations and with the minimum incurred the cost for the video

---

[1] For instance, YouTube allows viewers to choose their preferred spatial resolution.

stream provider? How can we learn the affinity of a user-defined inter-action with the heterogeneous services? How can we strike a balance between the incurred cost of heterogeneous services and the perfor-mance offered? and finally, how should the heterogeneity of the allo-cated resources be modified (i.e., reconfigured) with respect to the type and rate arriving video streaming requests?

3. *Specialized clouds for video streaming.*

So far, the idea of cloud has been in the form a centralized facility that can be used for diverse purposes. However, as clouds are evolving to be more service-oriented, their business model is shifting to specific pur-pose service providers. The reason being service consumers tend to pur-chase high-level services with certain QoE characteristics (e.g., watching video and paying in a pay-as-you-watch manner) as opposed to basic ser-vices such as dealing with VMs [139] and pay on an hourly basis. Another example can be purchasing transcoding or transmuxing service with cer-tain startup delay guaranteed.

In addition, viewers need flexible and diverse subscription services for video streaming services they receive that is not offered under current gen-eral purpose cloud services. For instance, some viewers may prefer to pay for video content in a pay-as-you-watch manner and some others may prefer monthly flat rate subscriptions. Offering such detailed facilities entails creating cloud providers dedicated to video streaming services. Creation of such clouds introduces multiple challenges, such as possible connections with basic clouds, specific purpose VM and containers, het-erogeneity of underlying hardware among many other challenges.

4. *Single streaming engine for both live- and VOD-streaming.*

Live-streams and VODs are structurally similar and also have similar computational demands. However, processing them is not entirely the same. In particular, live streaming tasks have a hard deadline and have to be dropped if they miss their deadline [12,140] whereas VOD tasks have a soft deadline and can tolerate deadline violations. Accordingly, in VOD, upon violating deadlines for a video segment, the deadlines of all segments behind it in the same stream should be updated. That is, VOD tasks have dynamic deadlines which is not the case for live streaming. In addition, there is no historic computational information for live streaming tasks to predict the execution time of new arriving tasks. This is not the case for VOD tasks.

Based on these differences, video stream providers utilize different video streaming engines and even different resources for processing and providing the services [12, 46]. The question is that how can we have an integrated streaming engine that can accommodate both of the streaming types simultaneously on a single set of allocated resources? How will this affect the scheduling and processing time estimation of the streaming engine?

5. *Blockchain technology for video streaming.*

The idea of blockchain is to create a secure peer-to-peer communication through a distributed ledger (database). In this system, every network node owns a copy of the blockchain, and every copy includes the full history of all recorded transactions. All nodes must agree on a transaction before it is placed in the blockchain.

The idea has rapidly got adopted and is being extensively developed in various domains to improve traceability and auditability [139]. We envisage that this technology will have a great impact in the video streaming industry. Some of the applications can be as follows: Whitelisting, which means keeping a list of legitimate publishers or distributors; Identity Management: the ability to perform identity management in a decentralized manner.

In addition, blockchain technology will grant more control and opportunity to video producers/publishers. In fact, the current video streaming industry is driven by quality expected and algorithms embedded in stream service providers. For instance, ordinary people cannot be publishers on Netflix. Even in the case of YouTube that enables ordinary users to publish content, the search and prioritization algorithms are driven by YouTube and not by the publisher. These limitations are removed in blockchain and publishers have more freedom in exposing their generated content. The secure distribution network of blockchain can be also useful for current streaming service providers (e.g., Netflix). They can use the network to securely maintain multiple versions of videos near viewers and distribute them with low latency and at the meantime reduce their cloud storage costs.

6. *Reuse and approximate computing for video streaming.*

Several video streaming viewers may request the same video under the same or dissimilar configurations (i.e., processing requirements). Current scheduling methods treat each video streaming request separately. That is, for each video streaming request, they have to be processed separately, even though the same video is being streamed simultaneously for two different viewers.

To make video streaming cost-efficient, for similar streaming requests (e.g., streaming the same video for two users with different resolutions) instead of repeatedly decoding the original video and then reencoding it, we can reuse processing and decreases the total processing time. For instance, the decoding of a video segment can be done once across all similar requests and then reencoding is performed separately. In addition to saving cost, this can potentially reduce congestion in scheduling queues by reducing the number of tasks waiting for processing and shortening the overall execution time. While this approach is interesting and there are preliminary efforts to address that (e.g., [121]), it can be challenging as integrating processing of the same video can jeopardize other video streaming tasks to be starved and viewers' QoE is impacted.

Research studies are needed to be undertaken and address this challenge. One question can be on how to integrate video streaming tasks from different viewers to make a more efficient use of cloud services? How does this approach impact other video streaming tasks? Also, some video streaming tasks might be semantically similar and with some approximation, the result of processing for one request can be used for another request. For instance, two resolutions can be close and compatible. Then, the question is how to identify semantically similar streaming tasks in the system?

7. *Machine learning for video processing.*

Video encoding/decoding requires multiple predictions, e.g., intra prediction and inter prediction [141]. Machine learning can play an important role to keep improving these predictions to produce smaller size video with the identical quality.

An accurate task execution time estimation can significantly benefit task scheduling and resource provisioning in cloud [14, 49]. However, predicting the time is not effortless. It is proven that there is an affinity between GOP size and certain video stream processing tasks (e.g., transcoding time). However, better estimation can be achieved by using machine learning approaches.

Deneke et al. [15] utilize a machine learning technique to predict each video segment's transcoding time before scheduling. It shows significantly better load balancing than classical methods. Accordingly, one future research direction in cloud-based video streaming is to use the machine learning techniques to enable unsupervised learning of video streaming tasks for user-defined interactions. The estimation can be different for heterogeneous cloud VMs because various video processing tasks have a different affinity with the heterogeneous VMs.

8. *Reliable cloud-based video stream processing.*

Reliability of a video streaming service is based on its tolerance against possible failures. Streaming service providers receive services under a certain service level agreement (SLA) with the cloud service provider. SLA explains the availability of services and the latency of accessing them. A video streaming engine translates the SLA terms to its service level objectives (SLO) [142] and attempts to respect them even in the presence of failures. Failures can be of two types: cloud service (e.g., VM or container) failures, and video streaming tasks failures.

- *Cloud service fault tolerance.* Cloud service (e.g., VM) availability is vital for streaming service providers. To maintain good availability, when one server fails, its workloads need to be migrated to another server to keep the streaming service uninterrupted. Service fault tolerance has been widely studied in cloud computing and solutions for that mainly include redundancy of cloud services and data checkpointing [143, 144].

- *Video processing fault tolerance.* Some video streaming tasks can fail during processing. Video streaming engines should include policies to cope with the failure of video streaming tasks dispatching to the scheduling queue. The policies can redispatch the failed task for VOD streams or ignore it for live streaming.

Currently, there is no failure-aware solution tailored for video streaming processing. Given the specific characteristics of video streaming services, in terms of large data-size, expensive computation, and unique QoE expectations, it will be appropriate to investigate failure-aware solutions for reliable video streaming service.

9. *Federation of edge clouds for low-latency video streaming.*

To efficiently serve customer demands around the world, cloud service providers setup datacenters in various geographical locations [145, 146]. For example, Netflix utilizes Open Connect [147] in numerous geographical locations to minimize latency of video streaming.

However, existing cloud-based video streaming systems do not fully take advantage of this large distributed system to improve quality and cost of streaming. Mechanisms and policies are required to dynamically coordinate load distribution between the geographically distributed data centers and determine the optimal datacenter to provide streaming service for each video (e.g., for storage, processing, or delivery).

To address this problem, Buyya et al. [145] advocate the idea of creating the federation of cloud environments. In the context of video streaming, a cost-efficient and low latency streaming can be achieved by federating edge datacenters and take advantage of cached contents or processing power of neighboring edge datacenters. Specifically, solutions are required to stream a video not only from the nearest datacenter (the way CDNs conventionally operate), but also from neighboring edge datacenters. Such solutions should consider the trade-off between the cost of processing a requested video on a local edge datacenter, and getting that from a nearby edge.

References

[1] Global Internet Phenomena Report, 2018. https://www.sandvine.com/hubfs/downloads/phenomena/2018-phenomena-report.pdf.
[2] Netflix Is Now Bigger Than Cable TV, 2017. https://www.forbes.com/sites/ianmorris/2017/06/13/netflix-is-now-bigger-than-cable-tv/. June.
[3] G. Suciu, M. Anwar, R. Mihalcioiu, Virtualized video and Cloud Computing for efficient elearning, in: Proceedings of The International Scientific Conference eLearning and Software for Education, vol. 2, 2017, p. 205.
[4] C.-W. Lin, Y.-C. Chen, M.-T. Sun, Dynamic region of interest transcoding for multipoint video conferencing, IEEE Trans. Circuits Syst. Video Technol. 13 (10) (2003) 982–992.
[5] R.T. Collins, A.J. Lipton, T. Kanade, H. Fujiyoshi, D. Duggins, Y. Tsin, D. Tolliver, N. Enomoto, O. Hasegawa, P. Burt, et al., A system for video surveillance and monitoring, Technical report CMU-RI-TR-00-12, Robotics Institute, Carnegie Mellon University, 2000, pp. 1–68.
[6] H. Joo, H. Song, D.-B. Lee, I. Lee, An effective IPTV channel control algorithm considering channel zapping time and network utilization, IEEE Trans. Broadcast. 54 (2) (2008) 208–216.
[7] X. Li, M.A. Salehi, M. Bayoumi, N.-F. Tzeng, R. Buyya, Cost-efficient and robust on-demand video transcoding using heterogeneous cloud services, IEEE Trans. Parallel Distrib. Syst. (TPDS) 29 (3) (2018) 556–571.
[8] J. He, D. Wu, Y. Zeng, X. Hei, Y. Wen, Toward optimal deployment of cloud-assisted video distribution services, IEEE Trans. Circuits Syst. Video Technol. 23 (10) (2013) 1717–1728.
[9] X. Li, M.A. Salehi, M. Bayoumi, R. Buyya, CVSS: a cost-efficient and QoS-aware video streaming using cloud services, in: Proceedings of the 16th ACM/IEEE International Conference on Cluster Cloud and Grid Computing, May, 2016.
[10] X. Cheng, J. Liu, C. Dale, Understanding the characteristics of internet short video sharing: a youtube-based measurement study, IEEE Trans. Multimedia 15 (5) (2013) 1184–1194.
[11] L.C.O. Miranda, R.L.T. Santos, A.H.F. Laender, Characterizing video access patterns in mainstream media portals, in: Proceedings of the 22nd International Conference on World Wide Web, ACM, 2013, pp. 1085–1092.
[12] X. Li, M.A. Salehi, M. Bayoumi, VLSC:video live streaming using cloud services, in: Proceedings of the 6th IEEE International Conference on Big Data and Cloud Computing Conference, October, 2016.

[13] Netflix and AWS, 2021. https://aws.amazon.com/solutions/case-studies/netflix/. (accessed February).

[14] M. Salehi, R. Buyya, Adapting market-oriented scheduling policies for cloud computing, in: Proceedings of the Algorithms and Architectures for Parallel Processing, January, 6081, 2010, pp. 351–362. vol.

[15] T. Deneke, H. Haile, S. Lafond, J. Lilius, Video transcoding time prediction for proactive load balancing, in: Proceedings of the IEEE International Conference on Multimedia and Expo, ICME '14, pp. 1–6.

[16] F. Jokhio, T. Deneke, S. Lafond, J. Lilius, Analysis of video segmentation for spatial resolution reduction video transcoding, in: Proceedings of the IEEE International Symposium on Intelligent Signal Processing and Communications Systems (ISPACS), December, 2011, pp. 1–6.

[17] S. Lin, X. Zhang, Q. Yu, H. Qi, S. Ma, Parallelizing video transcoding with load balancing on cloud computing, in: Proceedings of the IEEE International Symposium on Circuits and Systems (ISCAS), May, 2013, pp. 2864–2867.

[18] F. Jokhio, A. Ashraf, S. Lafond, I. Porres, J. Lilius, Prediction-based dynamic resource allocation for video transcoding in cloud computing, in: Proceedings of the 21st IEEE Euromicro International Conference on Parallel, Distributed and Network-Based Processing (PDP), February, 2013, pp. 254–261.

[19] F. Lao, X. Zhang, Z. Guo, Parallelizing video transcoding using map-reduce-based cloud computing, in: Proceedings of the IEEE International Symposium on Circuits and Systems, May, 2012, pp. 2905–2908.

[20] H. Schwarz, D. Marpe, T. Wiegand, Overview of the scalable video coding extension of the H. 264/AVC standard, IEEE Trans. Circuits Syst. Video Technol. 17 (9) (2007) 1103–1120.

[21] D. Mukherjee, J. Bankoski, A. Grange, J. Han, J. Koleszar, P. Wilkins, Y. Xu, R. Bultje, The latest open-source video codec VP9—an overview and preliminary results, in: Picture Coding Symposium (PCS), 2013, IEEE, 2013, pp. 390–393.

[22] G.J. Sullivan, J.-R. Ohm, W.-J. Han, T. Wiegand, Overview of the High Efficiency Video Coding (HEVC) standard, IEEE Trans. Circuits Syst. Video Technol. 22 (2012) 1649–1668. 12.

[23] C. Timmerer, C. Müller, HTTP streaming of MPEG media, in: Streaming Day, 2010.

[24] MPEG-2 Transport of Compressed Motion Imagery and Metadata, 2014. http://www.gwg.nga.mil/misb/docs/standards/ST1402.pdf.

[25] R. Fielding, J. Gettys, J. Mogul, H. Frystyk, L. Masinter, P. Leach, T. Berners-Lee, Hypertext Transfer Protocol-HTTP/1.1, 1999. Tech. rep.

[26] B. Lesser, Programming Flash Communication Server, O'Reilly Media, Inc., 2005.

[27] H. Schulzrinne, Real Time Streaming Protocol (RTSP), 1998.

[28] H.-N. Chen, M. Rutman, C.D. MacLean, E.C. Hiar, G.A. Morten, Progressive download or streaming of digital media securely through a localized container and communication protocol proxy, 2012. US Patent 8,243,924.

[29] R. Pantos, W. May, HTTP Live Streaming, 2017.

[30] T. Stockhammer, Dynamic adaptive streaming over HTTP: standards and design principles, in: Proceedings of the Second Annual ACM Conference on Multimedia Systems, ACM, 2011, pp. 133–144.

[31] S.S. Krishnan, R.K. Sitaraman, Video stream quality impacts viewer behavior: inferring causality using quasi-experimental designs, IEEE/ACM Trans. Netw. 21 (6) (2013) 2001–2014.

[32] R. Buyya, M. Pathan, A. Vakali, Content Delivery Networks, vol. 9, Springer Science & Business Media, 2008.

[33] N. Ramzan, H. Park, E. Izquierdo, Video streaming over P2P networks: challenges and opportunities, Signal Process. Image Commun. 27 (5) (2012) 401–411.

[34] Z. Zhuang, C. Guo, Building cloud-ready video transcoding system for content delivery networks (CDNs), in: Proceedings of the Global Communications Conference, GLOBECOM '12, 2012, pp. 2048–2053.

[35] Y. Liu, Y. Guo, C. Liang, A survey on peer-to-peer video streaming systems, Peer Peer Netw. Appl. 1 (1) (2008) 18–28.

[36] W. Zhang, S. Cheung, M. Chen, Hiding privacy information in video surveillance system, in: Proceedings of the IEEE International Conference on Image Processing, ICIP 2005, vol. 3, 2005, p. II-868.

[37] Q. Liu, R. Safavi-Naini, N.P. Sheppard, Digital rights management for content distribution, in: Proceedings of the Australasian Information Security Workshop Conference on ACSW Frontiers 2003—Volume 21, ACSW Frontiers '03, Australian Computer Society, Inc., Darlinghurst, Australia, Australia, 2003, pp. 49–58, ISBN: 1-920682-00-7.

[38] F.F.A. Quayed, S.S. Zaghloul, Analysis and evaluation of Internet Protocol Television (IPTV), in: Proceedings of the Third International Conference on e-Technologies and Networks for Development (ICeND2014), April, 2014, pp. 162–164.

[39] G. Erdemir, O. Selvi, V. Ertekin, G. Eşgi, Project PISCES: developing an in-flight entertainment system for smart devices, 2017.

[40] E. Latja, Parallel Acceleration of H.265 Video Processing (Ph.D. thesis), Aalto University, 2017.

[41] S. Jana, E. Baik, A. Pande, P. Mohapatra, Improving mobile video telephony, in: Proceedings of the 11th Annual IEEE International Conference on Sensing, Communication, and Networking, 2014, pp. 495–503.

[42] J. Jo, J. Kim, Synchronized one-to-many media streaming with adaptive playout control, in: Proceeding of the International Society for Optics and Photonics on Multimedia Systems and Applications V, 4861, 2002, pp. 71–83. vol.

[43] M. Hosseini, M. Amini Salehi, R. Gottumukkala, Enabling interactive video stream prioritization for public safety monitoring through effective batch scheduling, in: Proceedings of the 19th IEEE International Conference on High Performance Computing and Communications, Thailand, December, HPCC '17, 2017.

[44] G. Gualdi, A. Prati, R. Cucchiara, Video streaming for mobile video surveillance, IEEE Trans. Multimedia 10 (6) (2008) 1142–1154.

[45] M. Smole, Live-to-VoD Streaming, 2017. https://bitmovin.com/bitmovins-live-vod-service/.

[46] X. Li, M.A. Salehi, M. Bayoumi, High perform on-demand video transcoding using cloud services, in: Proceedings of the 16th ACM/IEEE International Conference on Cluster Cloud and Grid Computing, May, CCGrid '16, 2016.

[47] I. Ahmad, X. Wei, Y. Sun, Y.-Q. Zhang, Video transcoding: an overview of various techniques and research issues, IEEE Trans. Multimedia 7 (5) (2005) 793–804.

[48] A. Vetro, C. Christopoulos, H. Sun, Video transcoding architectures and techniques: an overview, IEEE Signal Process. Mag. 20 (2) (2003) 18–29.

[49] X. Li, M. Amini Salehi, Y. Joshi, M.K. Darwich, B. Landreneau, M. Bayoumi, Performance analysis and modeling of video transcoding using heterogeneous cloud services, IEEE Trans. Parallel Distrib. Syst. 30 (4) (2018) 910–922.

[50] O. Werner, Requantization for transcoding of MPEG-2 intraframes, IEEE Trans. Image Process. 8 (1999) 179–191.

[51] N. Bjork, C. Christopoulos, Transcoder architectures for video coding, IEEE Trans. Consum. Electron. 44 (1) (1998) 88–98.

[52] S. Goel, Y. Ismail, M. Bayoumi, High-speed motion estimation architecture for real-time video transmission, Comput. J. 55 (1) (2012) 35–46.

[53] B.G. Haskell, A. Puri, A.N. Netravali, Digital Video: An Introduction to MPEG-2, Springer Science & Business Media, 1996.

[54] T. Wiegand, G.J. Sullivan, G. Bjontegaard, A. Luthra, Overview of the H. 264/AVC video coding standard, IEEE Trans. Circuits Syst. Video Technol. 13 (7) (2003) 560–576.

[55] C.E. Jones, K.M. Sivalingam, P. Agrawal, J.C. Chen, A survey of energy efficient network protocols for wireless networks, Wireless Netw. 7 (4) (2001) 343–358.

[56] S. Balcisoy, M. Karczewicz, T. Capin, Progressive Downloading of Timed Multimedia Content, 2005. U.S. Patent Application 10/865,670, January 6.

[57] V.N. Padmanabhan, J.C. Mogul, Improving HTTP latency, Comput. Netw. ISDN Syst. 28 (1–2) (1995) 25–35.

[58] V. Jacobson, R. Frederick, S. Casner, H. Schulzrinne, RTP: a transport protocol for real-time applications, 2003.

[59] T.C. Thang, H.T. Le, A.T. Pham, Y.M. Ro, An evaluation of bitrate adaptation methods for HTTP Live Streaming, IEEE J. Sel. Areas Commun. 32 (4) (2014) 693–705.

[60] A. Begen, T. Akgul, M. Baugher, Watching video over the web: Part 1: Streaming protocols, IEEE Internet Comput. 15 (2) (2011) 54–63.

[61] N. Bouzakaria, C. Concolato, J. Le Feuvre, Overhead and performance of low latency live streaming using MPEG-DASH, in: Proceedings of the 5th International Conference on Information, Intelligence, Systems and Applications, 2014, pp. 92–97.

[62] C.M.A. Format, 2016. https://mpeg.chiariglione.org/standards/mpeg-a/common-media-application-format/text-isoiec-cd-23000-19-common-media-application. (June).

[63] C. Hanna, D. Gillies, E. Cochon, A. Dorner, J. Alred, M. Hinkle, Demultiplexer IC for MPEG2 transport streams, IEEE Trans. Consum. Electron. 41 (3) (1995) 699–706.

[64] S. Saroiu, K.P. Gummadi, R.J. Dunn, S.D. Gribble, H.M. Levy, An analysis of internet content delivery systems, SIGOPS Oper. Syst. Rev. 36 (SI) (2002) 315–327, https://doi.org/10.1145/844128.844158.

[65] A. Vakali, G. Pallis, Content delivery networks: status and trends, IEEE Internet Comput. 7 (6) (2003) 68–74, https://doi.org/10.1109/MIC.2003.1250586.

[66] V.K. Adhikari, Y. Guo, F. Hao, M. Varvello, V. Hilt, M. Steiner, Z.-L. Zhang, Unreeling Netflix: understanding and improving multi-CDN movie delivery, in: Proceedings of the IEEE International Conference on Computer Communications, INFOCOM '12, 2012, pp. 1620–1628.

[67] J. Apostolopoulos, T. Wong, W.-T. Tan, S. Wee, On multiple description streaming with content delivery networks, in: Proceedings of the 21st Annual Joint Conference of the IEEE Computer and Communications Societies, 3, 2002, pp. 1736–1745. vol.

[68] C.D. Cranor, M. Green, C. Kalmanek, D. Shur, S. Sibal, J.E. Van der Merwe, C.J. Sreenan, Enhanced streaming services in a content distribution network, IEEE Internet Comput. 5 (4) (2001) 66–75.

[69] S. Wee, J. Apostolopoulos, W.-T. Tan, S. Roy, Research and design of a mobile streaming media content delivery network, in: ICME '03, Proceedings of the International Conference on Multimedia and Expo, vol. 1, 2003. I–5.

[70] I. Benkacem, T. Taleb, M. Bagaa, H. Flinck, Performance benchmark of transcoding as a virtual network function in CDN as a service slicing, in: Proceedings of the IEEE Conference on Wireless Communications and Networking (WCNC), 2018.

[71] A.O. Al-Abbasi, V. Aggarwal, EdgeCache: an optimized algorithm for CDN-based over-the-top video streaming services, in: Proceedings of the IEEE Conference on Computer Communications Workshops (INFOCOM WKSHPS), IEEE, 2018, pp. 202–207.

[72] L. Kontothanassis, R. Sitaraman, J. Wein, D. Hong, R. Kleinberg, B. Mancuso, D. Shaw, D. Stodolsky, A transport layer for live streaming in a content delivery network, Proc. IEEE 92 (9) (2004) 1408–1419.

[73] H. Yin, X. Liu, T. Zhan, V. Sekar, F. Qiu, C. Lin, H. Zhang, B. Li, Design and deployment of a hybrid CDN-P2P system for live video streaming: experiences with LiveSky, in: Proceedings of the 17th ACM International Conference on Multimedia, ACM, 2009, pp. 25–34.

[74] Z.H. Lu, X.H. Gao, S.J. Huang, Y. Huang, Scalable and reliable live streaming service through coordinating CDN and P2P, in: Proceedings of the IEEE 17th International Conference on Parallel and Distributed Systems, ICPADS '11, 2011, pp. 581–588.

[75] M.M. Golchi, H. Motameni, Evaluation of the improved particle swarm optimization algorithm efficiency inward peer to peer video streaming, Comput. Netw. 142 (2018) 64–75.

[76] Y.-h. Chu, S.G. Rao, H. Zhang, A case for end system multicast (keynote address), in: ACM SIGMETRICS Performance Evaluation Review, vol. 28, ACM, 2000, pp. 1–12.

[77] J. Jannotti, D.K. Gifford, K.L. Johnson, M.F. Kaashoek, et al., Overcast: reliable multicasting with on overlay network, in: Proceedings of the 4th Conference on Symposium on Operating System Design & Implementation, vol. 4, USENIX Association, 2000, p. 14.

[78] N. Magharei, R. Rejaie, Y. Guo, Mesh or multiple-tree: a comparative study of live P2P streaming approaches, in: Proceedings of the 26th IEEE International Conference on Computer Communications, INFOCOM '07, 2007, pp. 1424–1432.

[79] V. Venkataraman, K. Yoshida, P. Francis, Chunkyspread: heterogeneous unstructured tree-based peer-to-peer multicast, in: Proceedings of the 2006 14th IEEE International Conference on Network Protocols, ICNP '06, 2006, pp. 2–11.

[80] S. Ahmad, C. Bouras, E. Buyukkaya, M. Dawood, R. Hamzaoui, V. Kapoulas, A. Papazois, G. Simon, Peer-to-peer live video streaming with rateless codes for massively multiplayer online games, Peer Peer Netw. Appl. 11 (1) (2018) 44–62.

[81] A. Vlavianos, M. Iliofotou, M. Faloutsos, BiToS: enhancing BitTorrent for supporting streaming applications, in: Proceedings of the 25th IEEE International Conference on Computer Communications, INFOCOM '06, 2006, pp. 1–6.

[82] Y. Guo, K. Suh, J. Kurose, D. Towsley, P2Cast: peer-to-peer patching scheme for VoD service, in: Proceedings of the 12th International Conference on World Wide Web, ACM, 2003, pp. 301–309.

[83] C. Xu, G.-M. Muntean, E. Fallon, A. Hanley, A balanced tree-based strategy for unstructured media distribution in P2P networks, in: Proceedings of the IEEE International Conference on Communications, ICC '08, 2008, pp. 1797–1801.

[84] W.-P.K. Yiu, X. Jin, S.-H.G. Chan, VMesh: distributed segment storage for peer-to-peer interactive video streaming, IEEE J. Sel. Areas Commun. 25 (9) (2007).

[85] B. Cheng, H. Jin, X. Liao, Supporting VCR functions in P2P VoD services using ring-assisted overlays, in: Proceedings of the IEEE International Conference on Communications, ICC '07, 2007, pp. 1698–1703.

[86] T. Xu, J. Chen, W. Li, S. Lu, Y. Guo, M. Hamdi, Supporting VCR-like operations in derivative tree-based P2P streaming systems, in: Proceedings of the IEEE International Conference on Communications, ICC '09, 2009, pp. 1–5.

[87] M.M. Afergan, F.T. Leighton, J.G. Parikh, Hybrid content delivery network (CDN) and peer-to-peer (P2P) network, 2012 (December).

[88] D. Xu, S.S. Kulkarni, C. Rosenberg, H.-K. Chai, Analysis of a CDN-P2P hybrid architecture for cost-effective streaming media distribution, Multimedia Syst. 11 (4) (2006) 383–399.

[89] F. Dufaux, T. Ebrahimi, Scrambling for privacy protection in video surveillance systems, IEEE Trans. Circuits Syst. Video Technol. 18 (8) (2008) 1168–1174.

[90] P. Carrillo, H. Kalva, S. Magliveras, Compression independent object encryption for ensuring privacy in video surveillance, in: Proceedings of the IEEE International Conference on Multimedia and Expo, 2008, 2008, pp. 273–276.

[91] E.M. Al-Hurani, H.R. Al-Zoubi, Mitigation of DOS attacks on video trafficin wireless networks for better QoS, in: Proceedings of the 8th International Conference on Computer Modeling and Simulation, Canberra, Australia, 2017, pp. 166–169.

[92] M.A. Alsmirat, I. Obaidat, Y. Jararweh, M. Al-Saleh, A security framework for cloud-based video surveillance system, Multimed. Tools Appl. 76 (21) (2017) 22787–22802.

[93] G. Fehér, Enhancing wireless video streaming using lightweight approximate authentication, in: Proceedings of the 2nd ACM International Workshop on Quality of Service & Security for Wireless and Mobile Networks, Terromolinos, Spain, Q2SWinet '06, 2006, pp. 9–16.

[94] T. Thomas, S. Emmanuel, A.V. Subramanyam, M.S. Kankanhalli, Joint watermarking scheme for multiparty multilevel DRM architecture, IEEE Trans. Inf. Forensics Secur. 1556-6013, 4 (4) (2009) 758–767, https://doi.org/10.1109/TIFS.2009.2033229.

[95] FairPlay Streaming, https://developer.apple.com/streaming/fps/ (accessed May).

[96] K. Kumar, DRM on Android, in: Proceedings of the Annual IEEE India Conference, INDICON, '17, 2013, pp. 1–6.

[97] D.K. Dorwin, Application-driven playback of offline encrypted content with unaware DRM module, 2015. US Patent 9,110,902.

[98] F. Dobrian, V. Sekar, A. Awan, I. Stoica, D. Joseph, A. Ganjam, J. Zhan, H. Zhang, Understanding the impact of video quality on user engagement, in: Proceedings of the ACM SIGCOMM Computer Communication Review, vol. 41, ACM, 2011, pp. 362–373.

[99] M. Cha, H. Kwak, P. Rodriguez, Y.-Y. Ahn, S. Moon, I tube, you tube, everybody tubes: analyzing the world's largest user generated content video system, in: Proceedings of the 7th ACM SIGCOMM Conference on Internet Measurement, ACM, 2007, pp. 1–14.

[100] M.E.J. Newman, Power laws, Pareto distributions and Zipf's law, Contemp. Phys. 46 (5) (2005) 323–351.

[101] M. Darwich, M.A. Salehi, E. Beyazit, M. Bayoumi, Cost efficient cloud-based video streaming based on quantifying video stream hotness, Comput. J. (2018) 1085–1092, https://doi.org/10.1093/comjnl/bxy057.

[102] N.M. Nasrabadi, Pattern recognition and machine learning, J. Electron. Imaging 16 (4) (2007) 049901.

[103] Y. LeCun, Y. Bengio, G. Hinton, Deep learning, Nature 521 (7553) (2015) 436.

[104] C.A. Gomez-Uribe, N. Hunt, The Netflix recommender system: algorithms, business value, and innovation, ACM Trans. Manag. Inf. Syst. (TMIS) 6 (4) (2016) 13.

[105] Z. Miao, A. Ortega, Scalable proxy caching of video under storage constraints, IEEE J. Sel. Areas Commun. 0733-8716, 20 (7) (2002) 1315–1327, https://doi.org/10.1109/JSAC.2002.802061.

[106] P.J. Shenoy, H.M. Vin, Efficient striping techniques for variable bit rate continuous media file servers, Perform. Eval. 38 (3) (1999) 175–199.

[107] D. Wu, Y.T. Hou, W. Zhu, Y.-Q. Zhang, J.M. Peha, Streaming video over the Internet: approaches and directions, IEEE Trans. Circuits Syst. Video Technol. 11 (3) (2001) 282–300, https://doi.org/10.1109/76.911156.

[108] S. Newman, Building Microservices: Designing Fine-Grained Systems, O'Reilly Media, Inc., 2015.

[109] J. Thönes, Microservices, IEEE Software 32 (1) (2015) 116. 116.

[110] L. Ao, L. Izhikevich, G.M. Voelker, G. Porter, Sprocket: a serverless video processing framework, in: Proceedings of the ACM Symposium on Cloud Computing, SoCC '18, 2018, pp. 263–274, ISBN: 978-1-4503-6011-1.

[111] L. Wang, M. Li, Y. Zhang, T. Ristenpart, M. Swift, Peeking behind the curtains of serverless platforms, in: 2018 USENIX Annual Technical Conference

(USENIX ATC 18), USENIX Association, 2018, pp. 133–146. https://www.usenix.org/conference/atc18/presentation/wang-liang.

[112] M. Darwich, E. Beyazit, M.A. Salehi, M. Bayoumi, Cost efficient repository management for cloud-based on-demand video streaming, in: Proceedings of 5th IEEE International Conference on Mobile Cloud Computing, Services, and Engineering, April, 2017, pp. 39–44, https://doi.org/10.1109/MobileCloud.2017.23.

[113] M. Kim, Y. Cui, S. Han, H. Lee, Towards efficient design and implementation of a Hadoop-based distributed video transcoding system in cloud computing environment, Int. J. Multimedia Ubiquit. Eng. 8 (2) (2013) 213–224.

[114] A. Ashraf, F. Jokhio, T. Deneke, S. Lafond, I. Porres, J. Lilius, Stream-based admission control and scheduling for video transcoding in cloud computing, in: Proceedings of the 13th IEEE/ACM International Symposium on Cluster, Cloud and Grid Computing, May, CCGrid '13, 2013, pp. 482–489.

[115] Z. Li, Y. Huang, G. Liu, F. Wang, Z.-L. Zhang, Y. Dai, Cloud transcoder: bridging the format and resolution gap between internet videos and mobile devices, in: Proceedings of the 22nd International Workshop on Network and Operating System Support for Digital Audio and Video, June, 2012, pp. 33–38.

[116] C.-F. Lai, H.-C. Chao, Y.-X. Lai, J. Wan, Cloud-assisted real-time transrating for HTTP Live Streaming, IEEE Wireless Commun. 20 (3) (2013) 62–70.

[117] T.C. Thang, Q.-D. Ho, J.W. Kang, A.T. Pham, Adaptive streaming of audiovisual content using MPEG DASH, IEEE Trans. Consum. Electron. 58 (1) (2012) 78–85.

[118] F. Jokhio, A. Ashraf, S. Lafond, J. Lilius, A computation and storage trade-off strategy for cost-efficient video transcoding in the cloud, in: Proceedings of the 39th EUROMICRO Conference on Software Engineering and Advanced Applications (SEAA), September, 2013, pp. 365–372.

[119] H. Zhao, Q. Zheng, W. Zhang, B. Du, Y. Chen, A version-aware computation and storage trade-off strategy for multi-version VoD systems in the cloud, in: Proceedings of IEEE Symposium on Computers and Communication (ISCC), July, 2015, pp. 943–948.

[120] O. Barais, J. Bourcier, Y.-D. Bromberg, C. Dion, Towards microservices architecture to transcode videos in the large at low costs, in: Proceedings of the International Conference on Telecommunications and Multimedia, 2016, pp. 1–6.

[121] C. Denninnart, M. Amini Salehi, A.N. Toosi, X. Li, Leveraging computational reuse for cost- and QoS-efficient task scheduling in clouds, in: Proceedings of the 16th International Conference on Service-Oriented Computing, November, ICSOC '18, 2018.

[122] C. Timmerer, D. Weinberger, M. Smole, R. Grandl, C. Müller, S. Lederer, Live transcoding and streaming-as-a-service with MPEG-DASH, in: Proceedings of the IEEE International Conference on Multimedia & Expo Workshops (ICMEW), June, IEEE, 2015, pp. 1–4.

[123] R.L. Myers, P. Ranganathan, I. Chvets, K. Pakulski, Methods and systems for real-time transmuxing of streaming media content, 2017. US Patent 9,712,887.

[124] J. Gorostegui, A. Martin, M. Zorrilla, I. Alvaro, J. Montalban, Broadcast delivery system for broadband media content, in: Proceedings of the IEEE International Symposium on Broadband Multimedia Systems and Broadcasting, BMSB '17, 2017, pp. 1–9.

[125] R. Hussain, M. Amini Salehi, A. Kovalenko, O. Semiari, S. Salehi, Robust resource allocation using edge computing for smart oil field, in: Proceedings of the 24th International Conference on Parallel and Distributed Processing Techniques and Applications, July, PDPTA '18, 2018, pp. 495–503.

[126] F. Chen, K. Guo, J. Lin, T. La Porta, Intra-cloud lightning: building CDNs in the cloud, in: Proceedings of the IEEE Conference INFOCOM, IEEE, 2012, pp. 433–441.

[127] H. Hu, Y. Wen, T.-S. Chua, J. Huang, W. Zhu, X. Li, Joint content replication and request routing for social video distribution over cloud CDN: a community clustering method, IEEE Trans. Circuits Syst. Video Technol. 26 (7) (2016) 1320–1333.

[128] H. Hu, Y. Wen, T.-S. Chua, Z. Wang, J. Huang, W. Zhu, D. Wu, Community based effective social video contents placement in cloud centric CDN network, in: Proceeding of the IEEE International Conference on Multimedia and Expo, Chengdu, China, ICME '14, 2014, pp. 1–6.

[129] Y. Jin, Y. Wen, G. Shi, G. Wang, A.V. Vasilakos, CoDaaS: an experimental cloud-centric content delivery platform for user-generated contents, in: Proceedings of the International Conference on Computing, Networking and Communications, ICNC '12, IEEE, 2012, pp. 934–938.

[130] H. Li, L. Zhong, J. Liu, B. Li, K. Xu, Cost-effective partial migration of VoD services to content clouds, in: Proceedings of the 2011 IEEE International Conference on Cloud Computing (CLOUD), 2011, pp. 203–210.

[131] Y. Li, Y. Shen, Y. Liu, Utilizing content delivery network in cloud computing, in: Proceeding of the IEEE International Conference on Computational Problem-Solving (ICCP), Leshan, China, 2012, pp. 137–143.

[132] J. Edstrom, D. Chen, Y. Gong, J. Wang, N. Gong, Data-pattern enabled self-recovery low-power storage system for big video data, IEEE Trans. Big Data 5 (1) (2017) 95–105, https://doi.org/10.1109/TBDATA.2017.2750699.

[133] D.A. Rodriguez-Silva, L. Adkinson-Orellana, F.J. Gonz'lezCastao, I. Armio-Franco, D. Gonz'lez-Martnez, Video surveillance based on Cloud storage, in: Proceedings of the IEEE Fifth International Conference on Cloud Computing, 2012, pp. 991–992, https://doi.org/10.1109/CLOUD.2012.44.

[134] G. Gao, W. Zhang, Y. Wen, Z. Wang, W. Zhu, Towards cost-efficient video transcoding in media cloud: insights learned from user viewing patterns, IEEE Trans. Multimedia 17 (8) (2015) 1286–1296.

[135] D.K. Krishnappa, M. Zink, R.K. Sitaraman, Optimizing the video transcoding workflow in content delivery networks, in: Proceedings of the 6th ACM Multimedia Systems Conference, ACM, 2015, pp. 37–48.

[136] M. Amini Salehi, X. Li, HLSaaS: high-level live video streaming as a service, in: Stream2016 Workshop organized by Department of Energy, Washington, DC, March, 2016.

[137] Y. Lin, H. Shen, CloudFog: leveraging Fog to extend cloud gaming for thin-client MMOG with high quality of service, IEEE Trans. Parallel Distrib. Syst. 28 (2) (2017) 431–445.

[138] D. Kumar, G. Baranwal, Z. Raza, D.P. Vidyarthi, A survey on spot pricing in cloud computing, J. Netw. Syst. Manag. 26 (4) (2018) 809–856.

[139] R. Buyya, S.N. Srirama, G. Casale, R. Calheiros, Y. Simmhan, B. Varghese, E. Gelenbe, et al., A Manifesto for future generation cloud computing: research directions for the next decade, ACM Comput. Surv. (CSUR) 51 (5) (2018) 1–38.

[140] A. Mokhtari, C. Denninnart, M.A. Salehi, Autonomous task dropping mechanism to achieve robustness in heterogeneous computing systems, in: 2020 IEEE International Parallel and Distributed Processing Symposium Workshops (IPDPSW), May 2020, pp. 17–26.

[141] R. Zhang, S.L. Regunathan, K. Rose, Video coding with optimal inter/intra-mode switching for packet loss resilience, IEEE J. Sel. Areas Commun. 18 (6) (2000) 966–976.

[142] G. Liu, H. Shen, H. Wang, An economical and SLO-guaranteed cloud storage service across multiple cloud service providers, IEEE Trans. Parallel Distrib. Syst. 28 (9) (2017) 2440–2453.

[143] S. Malik, F. Huet, Adaptive fault tolerance in real time cloud computing, in: Proceedings of the 2011 IEEE World Congress on Services, July, 2011, pp. 280–287.

[144] A. Bala, I. Chana, Fault tolerance-challenges, techniques and implementation in cloud computing, Int. J. Sci. Res. Publ. 9 (1) (2012). 1694-0814.

[145] M.A. Salehi, B. Javadi, R. Buyya, QoS and preemption aware scheduling in federated and virtualized Grid computing environments, J. Parallel Distrib. Comput. 72 (2) (2012) 231–245.

[146] M. Amini Salehi, B. Javadi, R. Buyya, Preemption-aware admission control in a virtualized grid federation, in: Proceedings of the 26th IEEE International Conference on Advanced Information Networking and Applications, 2012, pp. 854–861.

[147] T. Böttger, F. Cuadrado, G. Tyson, I. Castro, S. Uhlig, Open connect everywhere: a glimpse at the internet ecosystem through the lens of the Netflix CDN, ACM SIGCOMM Comput. Commun. Rev. 48 (1) (2018) 28–34.

About the authors

Dr. Xiangbo Li received his PhD degree in computer engineering from University of Louisiana at Lafayette in 2016. He was previously employed by Brightcove Inc. He is currently working as software engineer at Amazon AWS IVS. His areas of expertise include cloud-based video encoding, transcoding, packaging and serverside ad insertion.

Dr. Mahmoud Darwich earned his bachelor's degree in Electrical, Electronics and Communications Engineering from Jami'at Bayrut Al-Arabiya and his master's and doctorate degrees in Computer Engineering from the University of Louisiana at Lafayette. He is currently an assistant professor of computer science in the Mathematical and digital sciences department at the Bloomsburg University of Pennsylvania. His research interests include cloud computing, video streaming, and VLSI design.

Dr. Mohsen Amini Salehi received his PhD in Computing and Information Systems from Melbourne University, Australia, in 2012. He is currently an assistant professor and director of the High Performance Cloud Computing Laboratory, School of Computing and Informatics at University of Louisiana Lafayette, USA. His research focus is on different aspects of Distributed and Cloud Computing including heterogeneity, load balancing, virtualization, resource allocation, energy-efficiency, and security.

Dr. Magdy Bayoumi received the BSc and MSc degrees in Electrical Engineering from Cairo University, Egypt, the MSc degree in Computer Engineering from Washington University, St. Louis, and the PhD degree in Electrical Engineering from the University of Windsor, Ontario. He was the Vice President for Conferences of the IEEE Circuits and Systems (CAS) Society. He is the recipient of the 2009 IEEE Circuits and Systems Meritorious Service Award and the IEEE Circuits and Systems Society 2003 Education Award.

CHAPTER FIVE

User behavior-ensemble learning based improving QoE fairness in HTTP adaptive streaming over SDN approach

Tasnim Abar[a], Asma Ben Letaifa[b], and Sadok El Asmi[a]
[a]COSIM Lab, Carthage University, Higher School of Communications of Tunis, Tunis, Tunisia
[b]MEDIATRON LAB, SUPCOM Tunisia, Carthage University, Higher School of Communications of Tunis, Tunis, Tunisia

Contents

Abstract

Quality of Experience (QoE) for multimedia services is formed by a wealthy interaction between system, context and human factors. Whereas system and context variables are broadly inquired about, few research studies consider human variables as sources of orderly change. In other hand, Dynamic adaptive streaming over HTTP (DASH) has developed as an efficient innovation for video streaming but for a DASH framework, a most common case is that a limited server bandwidth is competed by multiusers. So, in order to enhance perceived QoE and guarantee fairness in wireless network, we propose in this paper to model Ensemble Learning approach based on user behavior factors to allocate the bandwidth collaboratively for multiusers and we use the

Advances in Computers, Volume 123
ISSN 0065-2458
https://doi.org/10.1016/bs.adcom.2021.01.004

MPEG-DASH technique to ameliorate the visual quality of each user. The performance evaluation is conducted by using network simulator and the obtained results show that our approach can give significant gain in terms of perceived QoE fairness.

1. Introduction

With the increasing growth of the demand of watching videos from anywhere and anytime, video streaming over wireless networks has been rapidly in progress in the past few years. Based on [1], it was predicted that streaming video services over wireless networks will occupy 79% of the total network traffic in 2022.

Although, the success of multimedia content is frequently related to the "perceived" quality of the content that the player uses and enjoys. This perception in general is dependent on the content type (motion, static) and available network parameters (bandwidth, delay, etc.). Service providers have the ability to stream the content at the right network environment (bandwidth, packet loss, etc.) to guarantee suitable user satisfaction. Whereas it might appear self-evident that nothing less than the most elevated conceivable parameters would fulfill users, there are events which require streaming at lower parameters. For case, within the example of mobile streaming where transfer speed and transmission shape a bottleneck [2]. Furthermore, research show that it isn't required to stream videos at high condition to satisfy users [3,4]. For example, it is possible that one user's perceived quality on a video sequence with a lower parameter setting (30fps, 480pg) is nearly equivalent to the other user's perceived quality on another sequence with higher parameters (20fpd, 720pg). We talk here about "user behavior" which has a great impact on user satisfaction.

The user behavior combines (1) the user profile such as age, gender and user interests; (2) user personality which is a series of \internal properties that report to overt behaviors [5]; and (3) user cultural representing the collective programming of the mind distinguishing the individuals of one country or category of people from others which hence leads to a broad tendency to lean toward certain states of a airs over others [6].

In other hand, one of the ultimate objectives in wireless multimedia networks is to afford a user-centric fair-share of network resources, so that the user QoE is maximized for all users existing in the network. There is a strong need to guarantee QoE fairness over diverse devices. Recent researches in adaptive video streaming [7] facilitate the dynamic adjustment of the streaming bitrate to reduce video stalling and buffering times, and ultimately enhance the overall user satisfaction. However, these types of video adaptation define considerable problems. First, they are unsteady and bursty in nature, particularly when competing with other video clients and related streams [8,9]. Besides, video adjustment happens based upon each client's point of view. This can be risky since the client has no information of other clients on the same network and regularly points to selfishly enhance its own QoE. Such egotistical behavior may potentially lead to declined QoE for different clients due to increased network congestion [8,9,10].

This paper introduces a novel approach to improve the *QoE fairness* in wireless multimedia networks based on SDN approach and DASH service using user behavior in order to estimate the needed resources of each user. In fact, the use of SDN paradigm permits us to advantage from its points of interest: the separation between control and data layer, and the network becomes simpler in its configuration and management. The control layer defines the brain of the network. It allows a dynamic resource management, control, routing and forwarding data, etc. In the other side, DASH service can easily reduce the playback interruptions and initial playback delays. Its cost is lower and it is easy to be deployed [11].

The rest of the paper is organized as follows: we present a brief state of the art where we introduce the SDN approach, DASH service, QoE influence factors and Machine Learning and Ensemble Learning in more details. In Section 3, we depict the contributions of our work which composed to two phases where are more detailed and involved. In Section 4, we discuss and analyze the obtained simulation results to evaluate the performance of our proposed approach. Finally, the conclusion is addressed in the last section.

2. State of the art
2.1 SDN approach to enhance QoE

SDN is an approach to network architecture that allows to intelligently and centrally control or "program" the network using software applications. It allows to the operators to manage whole network in a consistent and total way, whatever is subjacent network technology. SDN technology makes it possible to program the behavior of the network in a controlled and centralized manner through software applications using open APIs. By opening up traditionally closed network platforms and implementing a common SDN control layer, operators can manage the entire network and its equipment in a consistent fashion, regardless of the complexity of the underlying network technology.

Research works today are being dedicated to enable the SDN approach to automatize, coordinate, and manage the deployment and operation of network services. Also, many surveys highlight the SDN approach to address their problems and challenges: Authors in [12] present comprehensive survey on SDN in which they describe things from introduction to motivation for SDN, detail the difference between SDN and traditional networks and its roots. They also analyze the hardware infrastructure, southbound and northbound APIs, network layers, SDN controllers, network programming languages applications. In paper [13], authors focused on SDN approach and its challenges, issues, solutions. First, they provided and cover its basic model and software used to build a computer network with the help of software defined network mechanism then software tools used are listed, challenges and issues are described. Nathan et al. in [14] try to contribute to an overarching understanding of the common concepts and diverse approaches toward practical embodiments of Network Service Orchestration (NSO) which is a combination between SDN, NFV and Cloud Computing. This survey shows the importance of SDN in the improvement of network services.

In terms of ameliorating the QoE of network services, the manipulation of SDN approach facilitates the achievement of research goals. In fact, work done in [15] proposes a novel QoE-aware SDN enabled NFV architecture for controlling and managing Future Multimedia Applications on 5G systems to enhance the QoE of the delivered multimedia services by meeting quality application requirements (QoE). Besides, in [16], authors rely on SDN environment to highlights the effectiveness of the proposed collaborative approach which aims to study the effect of the frequency of information exchange between Over-The-Top (OTT) service and Internet Service Provider (ISP) on user satisfaction and the use of network resources to accurately predict the QoE. Then, work in [17] is based on SDN approach that enable framework for cloud-fog interoperation, aiming at optimizing QoE and network management. They describe the necessity of fog computing as an ideal complement instead of Cloud Computing that introduces low latency and it is not well suited to the provision of location-based services. Then they present how SDN controller can be exploited to bridge the interoperability of cloud and fog computing. Finally, in recent years, SDN approach is the most used to improve the performance of network providers and user satisfaction [18,19,20,21].

2.2 DASH to enhance QoE

The best practices of video streaming have evolved considerably since the introduction of HTML5 in 2008. To overcome the limitations of progressive download streaming, the major online video industries have created adaptive bitrate streaming proprietary formats such as HAS, HTTP progressive delivery, etc. In recent years, a new standard has emerged to replace these existing formats and unify the video streaming stream: MPEGDASH. Table 1 makes a brief comparison between different video streaming services.

MPEG-DASH [22] is a new standard used for video streaming. It presents the Media Presentation Concept (MPD), which is a structured set of video/audio content: A multimedia presentation consisting of a sequence of one or more consecutive periods that do not overlap. Each period comprises one or more representations of the same multimedia content and has a startup time allocated to it from the start of the media presentation. Each representation specifies a video quality profile consisting of several parameters such as bandwidth, encoding, and resolution. Also a representation

Table 1 Comparison between different video streaming services.

RSTP	– Communication protocol at the application level – Establish and control multimedia sessions – Does not carry data, must be associated with a transport protocol (RTP)
HTTP progressive delivery	– Process of broadcasting video content – No session is required – The video is cut into short segments of the same duration – The user cannot jump to a certain point in a video
HTTP streaming	– Similar to HTTP Progressive Delivery, with other features – Customer's ability to request a certain portion of a server video – Suitable for long videos
HAS	– Adjusts the quality of networks with low BW – Videos available on different coding schemes – Cutting into pieces of different sizes: the higher the BW, the bigger the size
DASH	– The user has the control of the delivery according to the state of the network – It's not a protocol or a codec – It's a way to offer a high quality video stream on the Internet

contains one or more segments, represented and located by their Universal Resource Locator (URL). The segments contain fragments of the actual video content. The MPEG DASH protocol specifies the MPD syntax and semantics, segment format, and delivery protocol (HTTP). Fortunately, it allows flexible configurations to implement different types of streaming services. The following parameters can be flexibly selected: the size and duration of the segments, the number of representations and the profile of each representation (bit rate, CODEC, container format, etc). Regarding the behavior of the client, it can decide in a flexible way when and how to download segments and choose an appropriate representation. The overview of the functioning of the DASH protocol is summarized in the following figure (Fig. 1).

In recent years, MPEG-DASH is constantly used in recent research fields where the main objective is to improve the quality of video streaming. MPEG-DASH has become popular for enhancement video quality models, this technique is an answer to the famous question of how to solve the two opposite phenomena, i.e., service quality and bandwidth limitations [23,24,25].

For example, in [26] authors are based on MPEG-DASH and Machine Learning (GMM-HMM) process to improving QoS/QoE performance and reducing power consumption and storage size. In other work, authors propose MulseDASH, which is a novel multiple sensorial media content delivery solutions based on MPEG-DASH. MulseDASH is designed based on Multi-sensorial Media Interaction to improve user QoE levels during the whole video playing [27]. Jan Willem et al. present a method to decrease the performance problems that exist in networks such as hotels, apartment complexes, and airports with a high number of DASH players, i.e., DASH suffers video freezes [28]. Jan Willem in other work [29] investigates how

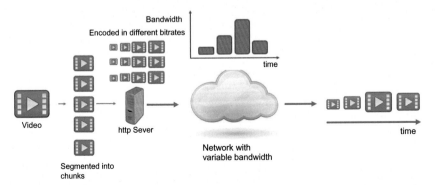

Fig. 1 How DASH works.

DASH Assisting Network Elements (DANEs) can help to enhance bottleneck links for video traffic, with the aim goal to better the user Experience QoE based on Markov Model. Last but not the least, Shuai Zhao et al. affirm with their research work in [30] that DASH can provide a smooth video quality experience and their approach also works steady within different network conditions.

2.3 Factors influencing the QoE

QoE indicates the total acceptability of an application or a service, such as the received subjectively by the end user [31].

Since the QoE is a subjective metric [31]. It is very hard to quantify and predict it. That's why today, service providers and researchers are facing new challenges related to measuring, and controlling the QoE. Then, investigating and analyzing the metrics that influence the QoE [32] is a first step to go.

Therefore, in order to evaluate the overall service quality, factors that influence the users' satisfaction should be determined in advance [33]. Table 2 resumes some research works from the literature where authors try to determine metrics that help them to model the user perception QoE.

Table 2 Metrics used to model the QoE.

References	Metrics used to model the QoE
[35]	Startup time, average bitrates, rebuffering
[36–38]	Variation of bitrates
[39]	Average bitrates, bitrate switches, Re-buffering ratio
[40,41]	Time elapsed since last rebuffering, Playback Indicator, Short Time Subjective Quality
[42]	User level, user behavior, pricing and charging, energy consumption
[43]	Join time, buffering ratio, rate of buffering events, average bitrate and rendering quality
[44]	Starvation Probability, Mean DT/VT Ratio
[45]	Start-up delay, buffer starvations when multiple video streams dynamically share the same base station
[46,47]	VQA metrics: PSNR, VQM, SSIM
[48]	PAcc (the possibility of a video quality making viewers pleasant or comfortable), GAcc (the possibility of a video quality being generally accepted by viewers)

Bref, Qualinet [34] has defined these factors as follows:

"Any characteristic of a user, system, service, application, or context whose actual state or setting may have influence on the Quality of Experience for the user."

In general, these influence factors can be classified into three categories (Table 2):

a. System factors

There are four classes that define the system factors indicated in Table 2: content which consist of video content and content reliability such as color, 2D/3D ..., media that defines all media configurations [49] (resolution, frame rate, etc.), network which refers to network metrics such as packet loss, delay ... and finally device factors [50] that include the equipment specifications, device capabilities, etc. Authors in [51] confirm that with the same video and same environment, the perceived QoE vary with the device characteristics.

b. Context factors

This category defines factors that take up the user's environment, in terms of physical (location and space), temporal, social (people present or involved in the experience), economic (Costs, type of subscription), task, and technical characteristics. In fact, when a user displays the same video on the same device, the perceived QoE can vary with the viewing environment (e.g., at home or on the street). Some literatures [52,53] argue that the home context may enhance the perceived quality with more efficient results compared with a standard laboratory or street context. Besides, Frohlich et al. [54] indicate that the quality perception and rating behavior change with the adjustment of video durations. They show that long clips (120 and 240s) are rated slightly higher than the short ones (10, 15 and 30s) especially for the videos of high visual qualities (Table 3). On the other hand, multitasking can degrade easily the user's perception. [55].

c. User behavior factors

While system and context factors are widely researched, few studies consider human or user behavior factors as QoE influence factors. Contrariwise these factors are main components that define QoE as somebody's individual experience. However, these metrics are highly complex and correlated to each other due to their subjectivity and relation to internal states and processes of human beings [56].

The user behavior factors include three classes:

(i) User profile which defines age, gender, interest [57]. For example, authors in [53] investigate that young adults intend to vote with lower QoE scores and woman prefer series movies than sport videos.

Table 3 Categories of QoE Influence factors.

Category	Class	Example
System factors	Content	2D/3D, color, texture, Spatial-temporal requirements
	Media	Bitrate, frame rate, resolution
	Network	Jitter, delay, packet loss, throughput, bandwidth
	Device	The equipment specifications, device capabilities
Context factor	Physical	Location, space, etc.
	Temporal	Playing time, duration, etc.
	Economic	Cost, brand, subscription type, etc.
	Task	Multitask, etc.
	Tech. characteristics	Relationship between the system of interest and other relevant systems
User behavior factors	User profile	Age, gender, interest
	User personality	Openness to experience, conscientiousness, extroversion, agreeableness and neuroticism
	User's cultural	Power distance; individualism; uncertainty avoidance; masculinity; pragmatism; and indulgence

(ii) User personality the personality of the user (openness to experience, conscientiousness, extroversion, agreeableness and neuroticism), has an essential impact to the QoE [58]. These features could map the user perception in many manners. For example, individuals with high neuroticism may be more sensitive to multimedia content that evokes negative emotions. [58]. In this work, we use the Five Factor Model (FFM) [59] to examine the predictive utility of personality.

(iii) User's cultural: this class defines the socio-cultural and educational background of the user. These dimensions play an important role in subjective QoE score [56,60]. For example, user with high indulgence can become more critical because of extended usage.

We define six levels of User's cultural in our work as presented in VSM-2013 questionnaire [61]: power distance; individualism; uncertainty avoidance; masculinity; pragmatism; and indulgence.

2.4 Machine learning and ensemble learning to model QoE

To attract or bind users to a service, real-time estimation of QoE is a necessity for network operators and service providers. It is therefore necessary to establish a correlation between QoE influence factors and the QoE, which can be used objectively, estimate the QoE. However, the relationship between these factors and QoE is difficult to estimate because it is not a linear relation. To do this, several models have been developed for the automatic assessment of user satisfaction; however, the accuracy of these models is still questionable. In addition, nowadays, researchers study human cognitive processes very closely and try to develop models that behave similarly to neurons in the human brain. Since the human mind is known to be non-deterministic, it is very difficult to get a formal algorithm of human behavior. This is the reason why researchers are turning to self-adaptive models and machine learning algorithms.

Today, several studies are based on the Ensemble Learning approach [62] in order to improve the performance of an intelligent prediction model. Indeed, Ensemble Learning technique is the combination of diverse Machine Learning models into one predictive model (same classifier or different classifier) Table 4.

Table 4 QoE Machine Learning and Ensemble learning Models.

References	Machine or ensemble learning	Algorithms	Results
[63]	Machine Learning	Support Vector Machine (SVM), Decision Tree (DT)	SVM: Mobile- correlation $88.59 \pm 2.85\%$, PDA-correlation $89.38 \pm 2.77\%$, Laptop-correlation $91.45 \pm 2.66\%$; – DT: Mobile-correlation $93.55 \pm 1.76\%$, PDA-correlation $90.29 \pm 2.61\%$ Laptop-correlation $95.46 \pm 2.09\%$
[64]	Machine Learning	Random Neural Networks	0.89
[65]	Machine Learning	Decision tree, Neural network, Bayesian	J48: correlation 0.98 MLP: correlation 0.92 Naives: correlation 0.78.

Table 4 QoE Machine Learning and Ensemble learning Models.—cont'd

References	Machine or ensemble learning	Algorithms	Results
[66]	Ensemble Learning	DT, boosting, bagging, stacking	DT is the most accurate single model to predict QoE, Ensemble learning models, and in particular stacking ones, are capable to significantly increase accuracy prediction and overall classification performance
[67]	Ensemble Learning	Boosting Support Vector Regression (SVM)	RMSE 0.49
[68]	Machine and Ensemble Learning	SVM, KNN, DT, Bagging, Stacking, RF	The stacking method obtains the best performance results with an overall accuracy of around 90%
[62]	Machine and Ensemble Learning	Bagging, boosting, stacking, DT, LWL NaivesBayesUpdateable	The stacking method obtains the best performance results with an overall accuracy of around 0.938

In the following table, we present some of the models that use Machine and Ensemble Learning algorithms for estimation of QoE.

3. Contributions

Multi-user video streaming case can be seen everywhere today as the video service becomes the "killer" service over the Internet [69]. The QoE fairness is a critical subject for researchers, content providers and ISP. Consequently, a multi-user QoE fairness approach of video streaming is of incredible significance.

The main objective of this work is to propose a novel framework based on SDN approach and DASH service to monitor the QoE of active users who are interested in using streaming video service. Our contributions can be summarized in two phases:

1. Phase I: Fairness bitrate allocation: the goal of this phase is to guarantee the satisfaction of all active users playing video streaming based on their behaviors (detailed in the previous section).

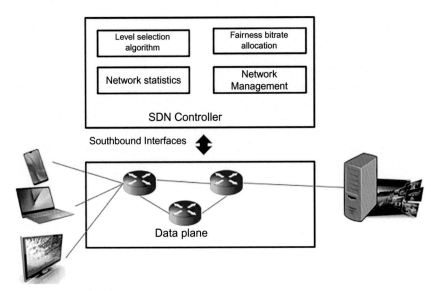

Fig. 2 Proposed architecture.

2. Phase II: Improvement of visual quality for each user: in this phase, we study how DASH interacts to guarantee reliable video quality with the minimum switching between levels. To do this, we are based on level selection algorithm proposed in [50].

Fig. 2 presents in more details our proposed architecture and how we integrate the SDN approach to our solution. Indeed, the SDN controller is responsible for network management and administration and the execution of level selection and fairness bitrate allocation solutions.

3.1 Phase I: Fairness bitrate allocation

Fig. 3 depicts the proposed approach to ensure the fairness bitrate allocation.

In fact, we are based on Ensemble learning algorithms to estimate the resources needed for each user based on its behavior. First, the SDN controller determines the active users who intend to watch a video streaming based on the Poisson process [50]. After that, we turn the ensemble learning model to predict the suitable bitrate with the request user behavior which guarantee an acceptable QoE.

(a) Poisson process

A Poisson process is a mathematical model modeling random events that recur over time: births, breakdowns, radioactive decay, etc. In our work, we use this process to model the number of active users in the network

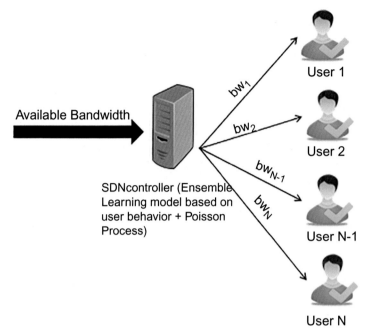

Fig. 3 Fairness bitrate allocation approach.

at time t. The number of clients N (t) arriving in the network during any time interval t follows a Poisson distribution of mean λt.

$$P(N(t) = k) = e^{-\lambda t} \frac{\lambda t^k}{k!} \tag{1}$$

λ: is a real positive constant.

In our work, the SDN controller uses this constant to distinguish the type of application used by a new client to distinguish which clients intending to use video streaming service. So we suppose that:

λ_1: For users who use streaming video: video conference, online video, etc.

λ_2: For users who use voice service such as audio call.

λ_3: For simple services such as downloading a pdf doc, facebook support, web service, etc.

In the rest of the work we are only interested in λ_1.

(b) Ensemble Learning model based on user behavior

In this part, we investigate our predictive model in order to estimate the suitable bitrate for each active user in the network which ensure its satisfaction based on its behavior. We start by detailing the dataset, after that we detail the steps to build our predictive model.

i. Dataset description

In this paper, we've chosen to inspire our dataset from the dataset published in [58], and not to rebuild a new dataset, due to the fact that, it is gathered using crowdsourcing which build a realistic environment that is highly cost-efficient, avoid biases and flexible for carrying user experiment, besides it contains the features we need to conduct our study. The authors have collected a set of 144 video sequences (from 12 short movie excerpts) were rated by 114 participants from a cross-cultural population, producing 1232 ratings. In this work [58] participants responded to several measurement scales: perception of quality; enjoyment; culture; and personality Table 5.

After refining this dataset by eliminating some features and adding the user profile features the table below details the features used in our dataset.

ii. Construction of Ensemble Learning Model

In this section, we investigate the construction of our predictive model based on Ensemble Learning technique. We use three Ensemble Learning algorithms defined as bellow:

1. Bagging: Given a sample of dataset, multiple bootstrapped subsamples are pulled. A single classifier is formed on each of the bootstrapped

Table 5 Details about features used in our dataset.

	Age
	Gender
User profile	Interest
User personality	Openness to experience
	Conscientiousness
	Extroversion
	Agreeableness
	Neuroticism
User's cultural	Power distance
	Individualism
	Uncertainty avoidance
	Masculinity
	Pragmatism
	Indulgence

subsamples. After each subsample classifier has been formed, an algorithm is utilized to aggregate over the classifier results to build the most efficient predictor. Example Random Forest.

2. Boosting: is a sequential process, where each subsequent model tries to correct the errors of the previous model. The succeeding models are dependent on the previous model.

3. Stacking: is an Ensemble Learning technique that uses predictions from multiple models (for example, decision tree, knn or svm, etc.) to form a new efficient predictive model.

We consider Recall and Precision as performance metrics presented in Eqs. 2 and 3:

$$Recall_i = \frac{TP_i}{n_i} \tag{2}$$

$$Precision_i = \frac{TP_i}{TP_i + FP_i} \tag{3}$$

Where TP_i is the number of occurrences where is accurately classified and n_i is the number of occurrences in class i. The precision is the percentage of occurrences correctly classified belonging to class i among all the occurrences classified as belonging to class i, including true and false positives FP_i.

3.2 Phase II: Improvement of visual quality for each user

In this phase, we are based on our previous work published in [50]. So, to improve the perceived quality for each user after determining the suitable bitrate allocation based on it behavior in the first phase, we use now two interconnected sub phases: (1) Machine Learning prediction sub phase to determine the most suitable level presented in Table 3 with the existing network conditions Table 6.

Table 6 Levels.

Level	Bitrates	Frame rate	Resolution
1	768 kbps	10 fps	512 × 240
2	1000 kbps	12 fps	640 × 360
3	2500 kbps	15 fps	848 × 480
4	3000 kbps	30 fps	720 × 540
5	3500 kbps	30 fps	1280 × 720
6	5500 kbps	50 fps	1920 × 1080
7	7000 kbps	50 fps	1920 × 1080

This level is considered as a threshold for the quality level selection of the current segment. To do this, we must involve the maximum number of intermediate segments to be downloaded defined in Algorithm 1.

Algorithm 1. Level selection algorithm.

1: Boolean change = false
2: Form: 2: M do
3: if change ==true then
4: $l_{best,m} = ML - Estimation()$
5: if $l_{best,m} > l_{m-1}$ then
6: $k = -1$
7: do
8: $k = k + 1$
9: $t_{buffer} = \dfrac{B(t) - B_{min}}{1 - \dfrac{b(l_{best,m})}{b(l_{best,m-k})}}$
10: $n_k = \dfrac{t_{buffer} * b(l_{best,m})}{\tau * b(l_{best,m-k})}$
11: while $(n_k < 1)$ && $(k < l_{best,m} - l_{m-1})$
12: end while
13: $l_m = l_{best,m} - k$
14: else
15: $k = -1$
16: do
17: $k = k + 1$
18: $t_{buffer} = \dfrac{B(t) - B_{min}}{1 - \dfrac{b(l_{best,m})}{b(l_{best,m+k})}}$
19: $n_k = \dfrac{t_{buffer} * b(l_{best,m})}{\tau * b(l_{best,m+k})}$
20: while $(n_k < 1)$ && $(k < l_{m-1} - 1)$
21: end while
22: $l_m = l_{best,m} - k$
23: end if
24: end if
25: end for

The Machine learning process is based on the available network resources, the type of connected device, and the video content (in motion, static, etc.) as input features. We note that the use of Machine Learning approach helps us to minimize the computations time and the complexity and facilitate the work [70]. (2) The second sub phase is based on Algorithm 1 where we compare at each segment between the predicted level and the level of previous segment. The goal here is to minimize the difference between levels of two successive segments in order to assure an acceptable visual quality.

4. Analysis and discuss
4.1 Ensemble learning classification model

In this part of simulation results, we use three Ensemble Learning algorithms simulated in Weka tool [71]: Random Forest as Bagging method, AdaBoost as Boosting method and for Stacking method, we combine three simple learning algorithms: Decision tree, NaivesBayesUpdateable and K-Nearest neighbors.

We investigate the impact of the variation of training set based on k-fold cross validation approach. The k varies from 8 to 15. The performances, Precision and Recall, are depicted in Fig. 4A and B, respectively.

From these two curves, we can observe that: first, each algorithm reaches its highest performance when k-cross-validation is 10 value. For that, we will consider this value for the rest of the experiment. Second, we can remark that Stacking and Random Forest models achieve high recall and precision but Stacking outperforms Bagging model ($0.95 > 0.92$ as recall metric). On the other hand, AdaBoost obtains the wrong results and fails to build performant model based on our dataset with recall and precision around 0.56.

So as a result, we can conclude that Stacking model is the best model for our approach to predict the suitable resources based on the user behavior with recall and precision around 0.92.

To highlight our results, Table 7 depicts the corresponding AUC values. AUC metric is a performance measurement for classification problem at various thresholds settings. It represents the degree or the measure of separability. It tells how much model is capable of distinguishing between classes. So, higher the AUC, better the model is at predicting the class.

We can remark from this table that Stacking approach achieves the highest performance results 0.956, even softly outperforming Random Forest method where AdaBoost approach fails to predict the suitable resources for each user (0.426) due to its low intensity.

4.2 Level stability analysis

In order to effectively evaluate the performance of our proposed approach, we run a number of experiments upon our testbed using two approaches:

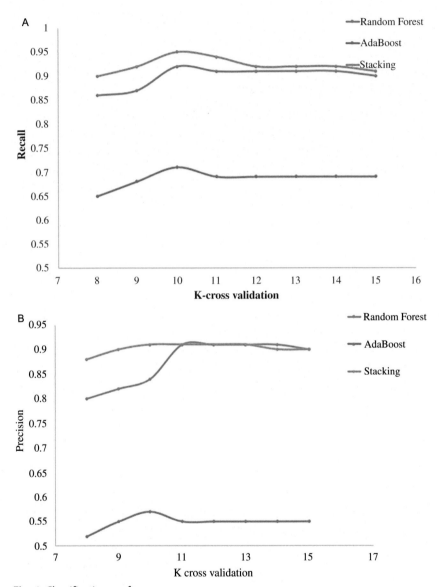

Fig. 4 Classification performances.

Table 7 ROC AUC results.

Random forest	0.896
AdaBoost	0.426
Stacking	0.956

Table 8 Characteristics of the videos.

	Duration	Type	Static or in motion	Resolution	Frame rate
1	180 s	Football	Motion	1920×1080	500fps
2	240 s	interview	static	1280×720	30fps

Table 9 User behaviors of the participants.

User	Age	Gender	Interest of	Personality	Cultural
1	35	Man	Video 1	Neuroticism	Power distance
2	30	Women	Video 2	Extroversion	Power distance
3	26	Women	No interest	Openness to experience	Indulgence

(1) the proposed method and (2) the allocation of the available bandwidth equally between connected users (using video service). Table 8 details the characteristics of the two videos used in our experience and Table 9 presents in details the three selected user behaviors for three participants.

We suppose the following scenario: user 1 starts playing video 1 at time 0 s. User 2 starts watching the video 2 at 60s where user 3 selects to watch video 3 at 120s. For each experiment, we measure the bitrates and the associated quality level.

Fig. 5 depicts the level of the downloaded video segments of the proposed scenario. Fig 5A presents the case when we use bandwidth equally approach and Fig. 5B presents our proposed approach. We note that bandwidth fairness is unaware of users' needs just it divides the available bandwidth according to the active users which causes the degradation of the QoE. In both cases user 1 benefits a good quality level. However, with the method of bandwidth equality, when user 2 starts watching the selected video, the available bandwidth is divided between the two clients and the quality of user 1 degrades and user 2 has bad quality from the beginning (both sessions start with a stall during 10s). With the intervention of user 3, user 1 sacrifices the most in competition over network resources while the other two clients seem to be unsatisfied with the poor quality.

While using our approach, users seem to be content with the perceived quality; we also notice that there is no stall or start delay thanks to our model which predict for each connected user the suitable resources that ensure his satisfaction.

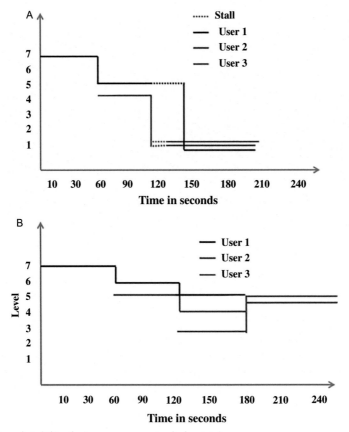

Fig. 5 Level stability (heterogeneous user behaviors).

5. Conclusion

In this paper, we propose a novel approach to improve the QoE fairness in wireless networks for multimedia service. This method involves on the principle of DASH, Ensemble learning technique and the advantages of SDN controller to optimize the user perception based on the user behavior. Testbed-based evaluations show that the proposed method provides stability and achieves to enhance the user satisfaction across heterogeneous user behaviors. As future work, we aim to build an automatic monitoring system that considers all QoE influence factors detailed in Table 3. Also, we look to use Information Centric Networking (ICN) approach which has an important role to enhance QoE.

References

[1] Cisco Corporation, Cisco VisualNnetworking Index: Global Mobile Data Traffic Forecast Update, 2017-2022, Cisco Corporation, 2019. Technology Report.

[2] W. Song, D. Tjondronegoro, M. Docherty, Saving bitrate vs. pleasing users: where is the break-even point in mobile video quality? in: Proceedings of the 19th ACM International Conference on Multimedia, 2011, pp. 403–412.

[3] G. Ghinea, J.P. Thomas, QoS impact on user perception and understanding of multimedia video clips, in: Proceedings of the Sixth ACM International Conference on Multimedia, 1998, pp. 49–54.

[4] V. Q. E. Group, et al., Final Report From the Video Quality Experts Group on the Validation of Objective Models of Video Quality Assessment, Phase ii (fr tv2), 2003. ftp://ftp.its.bldrdoc.gov/dist/ituvidq/BoulderVQEGjan04/VQEG PhaseII FRTVFinalReportSG9060E.doc.

[5] G. Matthews, I.J. Deary, M.C. Whiteman, Personality Traits, Cambridge University Press, 2003.

[6] G. Hofstede, Dimensionalizing cultures: the Hofstede model in context, Online Read. Psychol. Cult. 2 (1) (2011) 2307–0919.1014.

[7] ISO, ISO-IEC 23009–1:2012 Information Technology. Dynamic Adaptive Streaming over HTTP (DASH), 2012.

[8] S. Akhshabi, L. Anantakrishnan, A.C. Begen, et al., What happens when HTTP adaptive streaming players compete for bandwidth? in: Proceedings of the 22nd International Workshop on Network and Operating System Support for Digital Audio and Video, 2012, pp. 9–14.

[9] J. Esteban, S.A. Benno, A. Beck, et al., Interactions between HTTP adaptive streaming and TCP, in: Proceedings of the 22nd international workshop on Network and Operating System Support for Digital Audio and Video, 2012, pp. 21–26.

[10] T.-Y. Huang, N. Handigol, B. Heller, et al., Confused, timid, and unstable: picking a video streaming rate is hard, in: Proceedings of the 2012 Internet Measurement Conference, 2012, pp. 225–238.

[11] T. Stockhammer, Dynamic adaptive Sstreaming over HTTP— standards and design principles, in: Proceedings of the Second Annual ACM Conference on Multimedia Systems, 2011, pp. 133–144.

[12] D. Kreutz, F.M.V. Ramos, P.E. Verissimo, et al., Software-defined networking: a comprehensive survey, Proc. IEEE 103 (1) (2014) 14–76.

[13] D.S. Rana, S.A. Dhondiyal, S.K. Chamoli, Software defined networking (SDN) challenges, issues and solution, Int. J. Comput. Sci. Eng. 7 (2019) 1–7.

[14] N.F.S. De Sousa, D.A.L. Perez, R.V. Rosa, et al., Network service orchestration: a survey, Comput. Commun. 142 (2019) 69–94.

[15] A.A. Barakabitze, L. Sun, I.-H. Mkwawa, et al., A Novel QoE-Aware SDN-Enabled, NFV-Based Management Architecture for Future Multimedia Applications on 5G Systems, arXiv preprint arXiv:1904.09917, 2019.

[16] A. Ahmad, A. Floris, L. Atzori, Towards information-centric collaborative QoE management using SDN, in: 2019 IEEE Wireless Communications and Networking Conference (WCNC), IEEE, 2019, pp. 1–6.

[17] P. Yang, N. Zhang, Y. Bi, et al., Catalyzing cloud-fog interoperation in 5G wireless networks: an SDN approach, IEEE Netw. 31 (5) (2017) 14–20.

[18] Y. Bi, G. Han, C. Lin, et al., Mobility support for fog computing: an SDN approach, IEEE Commun. Mag. 56 (5) (2018) 53–59.

[19] M. Karakus, A. Durresi, A survey: control plane scalability issues and approaches in software-defined networking (SDN), Comput. Netw. 112 (2017) 279–293.

[20] L. Guillen, S. Izumi, T. Abe, et al., SDN implementation of multipath discovery to improve network performance in distributed storage systems, in: 2017 13th

International Conference on Network and Service Management (CNSM), IEEE, 2017, pp. 1–4.

[21] M.I. Lali, R.U. Mustafa, F. Ahsan, et al., Performance evaluation of software defined networking vs. traditional networks, Nucleus 54 (1) (2017) 6–22.

[22] MPEG, I., Information technology-dynamic adaptive streaming over http (dash)-part 1: Media presentation description and segment formats, ISO/IEC MPEG, Tech. Rep, 2012.

[23] J. Klink, M. Pasławski, P. Pawłowski, et al., Video quality assessment in the DASH technique, in: 2019 International Conference on Software, Telecommunications and Computer Networks (SoftCOM), IEEE, 2019, pp. 1–5.

[24] L. Zou, T. Bi, G.-M. Muntean, A DASH-based adaptive multiple sensorial content delivery solution for improved user quality of experience, IEEE Access 7 (2019) 89172–89187.

[25] T.C. Thang, Q.-D. Ho, J.W. Kang, et al., Adaptive streaming of audiovisual content using MPEG DASH, IEEE Trans. Consum. Electron. 58 (1) (2012) 78–85.

[26] K. Kanai, B. Wei, Z. Cheng, et al., Methods for adaptive video streaming and picture quality assessment to improve QoS/QoE performances, IEICE Trans. Commun. 102 (7) (2019) 1240–1247.

[27] L. Zou, T. Bi, G.-M. Muntean, A DASH-based adaptive multiple sensorial content delivery solution for improved user quality of experience, IEEE Access 7 (2019) 89172–89187.

[28] J.W. Kleinrouweler, B. Meixner, P. Cesar, Improving video quality in crowded networks using a DANE, in: Proceedings of the 27th Workshop on Network and Operating Systems Support for Digital Audio and Video, ACM, 2017.

[29] J.W. Kleinrouweler, Enhancing over-the-top video streaming quality with DASH assisting network elements, in: Adjunct Publication of the 2017 ACM International Conference on Interactive Experiences for TV and Online Video, ACM, 2017.

[30] S. Zhao, et al., Study of user QoE improvement for dynamic adaptive streaming over HTTP (MPEG-DASH), in: 2017 International Conference on Computing, Networking and Communications (ICNC), IEEE, 2017.

[31] UNION, I. T., ITU-T Recommendation P.910, "Subjective Video Quality Assessment Methods for Multimedia Applications", ITU, September 2008.

[32] J. Baraković Husić, S. Baraković, E. Cero, et al., Quality of experience for unified communications: a survey, Int. J. Netw. Manag. 30 (3) (2020), e2083.

[33] A. Mellouk, H.A. Tran, S. Hoceini, Quality-of-Experience for Multimedia, John Wiley & Sons, 2013.

[34] K. Brunnström, et al., Qualinet white paper on definitions of quality of experience, in: Eur. Netw. Qual. Exper. Multimedia Syst. Services, Lausanne, Switzerland, Tech. Rep., 2013. ffhal-00977812f. [Online]. Available https://hal.archives-ouvertes.fr/hal-00977812/document.

[35] I. Triki, Q. Zhu, R. Elazouzi, M. Haddad, Z. Xu, Learning from experience: a dynamic closed-loop QoE optimization for video adaptation and delivery, Comput. Sci. (2017).

[36] F. Dobrian, V. Sekar, A. Awan, I. Stoica, D. Joseph, A. Ganjam, J. Zhan, H. Zhang, Understanding the impact of video quality on user engagement, in: ACM SIGCOMM Computer Communication Review, no. 4, ACM, 2011.

[37] P. Juluri, V. Tamarapalli, D. Medhi, Measurement of quality of experience of video-on-demand services: a survey, IEEE Commun. Surv. Tutorials 18 (1) (2015) 401–418.

[38] P. Juluri, V. Tamarapalli, D. Medhi, Sara: segment aware rate adaptation algorithm for dynamic adaptive treaming over http, in: IEEE International Conference on Communication Workshop (ICCW), 2015.

[39] T. Mangla, E. Halepovic, M. Ammar, et al., Using session Modeling to estimate HTTP-based video QoE metrics from encrypted network traffic, IEEE Trans. Netw. Serv. Manag. 16 (3) (2019) 1086–1099.

[40] N. Eswara, et al., Streaming video QoE modeling and prediction: a long short-term memory approach, IEEE Trans. Circuits Syst. Video Technol. 30 (3) (2019) 661–673.

[41] N. Eswara, et al., Modeling continuous video qoe evolution: a state space approach, in: 2018 IEEE International Conference on Multimedia and Expo (ICME), IEEE, 2018.

[42] T. HoBfeld, M. Varela, P.E. Heegaard, L. Skorin-Kapov, Observations on emerging aspects in QoE modeling and their impact on QoE management, in: 2018 Tenth International Conference on Quality of Multimedia Experience (QoMEX), IEEE, 2018, pp. 1–6.

[43] F. Dobrian, V. Sekar, et al., Understanding the impact of video quality on user engagement, in: Proc. of ACM SIGCOMM 2011, vol. 41, 2011, pp. 362–373. no. 4.

[44] Y. Xu, et al., Modeling QoE of Video Streaming in Wireless Networks with Large-Scale Measurement of User Behavior, arXiv preprint arXiv:1604.01675, 2016.

[45] Y. Xu, S.E. Elayoubi, E. Altman, R. El-Azouzi, Impact of flowlevel dynamics on QoE of video streaming in wireless networks, in: 2013 Proceedings IEEE INFOCOM, IEEE, 2013, pp. 2715–2723. Turin, Italy.

[46] A. Eichhorn, P. Ni, Pick your layers wisely—a quality assessment of H. 264 scalable video coding for mobile devices, in: Presented at the IEEE International Conference on Communications, 2009. Dresden, Geremany.

[47] M.H. Pinson, S. Wolf, A new standardized method for objectively measuring video quality, IEEE Trans. Broadcast. 50 (2004) 312–322. Sep. 2004.

[48] W. Song, D.W. Tjondronegoro, Acceptability-based QoE models for mobile video, IEEE Trans. Multimedia 16 (3) (2014) 738–750.

[49] T. Abar, A. Ben Letaifa, S. El Asmi, Objective and subjective measurement QOE in SDN networks, in: 2017 13th International Wireless Communications and Mobile Computing Conference (IWCMC), IEEE, 2017, pp. 1401–1406.

[50] T. Abar, A. Ben Letaifa, S. El Asmi, Heterogeneous multiuser QoE enhancement over DASH in SDN networks, Wirel. Pers. Commun. 114 (4) (2020) 2975–3001.

[51] T. Abar, A. Ben Letaifa, S. Elasmi, Enhancing QoE based on machine learning and DASH in SDN networks, in: 2018 32nd International Conference on Advanced Information Networking and Applications Workshops (WAINA), IEEE, 2018, pp. 258–263.

[52] B. Gardlo, M. Ries, M. Rupp, et al., A QoE evaluation methodology for HD video streaming using social networking, in: 2011 IEEE International Symposium on Multimedia, IEEE, 2011, pp. 222–227.

[53] H.G. Msakni, H. Youssef, Is QoE estimation based on QoS parameters sufficient for video quality assessment? in: 2013 9th International Wireless Communications and Mobile Computing Conference (IWCMC), IEEE, 2013, pp. 538–544.

[54] P. Fröhlich, S. Egger, R. Schatz, et al., QoE in 10 seconds: are short video clip lengths sufficient for quality of experience assessment? in: 2012 Fourth International Workshop on Quality of Multimedia Experience, IEEE, 2012, pp. 242–247.

[55] J. Xue, C.W. Chen, A study on perception of mobile video with surrounding contextual influences, in: 2012 Fourth International Workshop on Quality of Multimedia Experience, IEEE, 2012, pp. 248–253.

[56] U. Reiter, K. Brunnström, K. De Moor, et al., Factors influencing quality of experience, in: Quality of Experience, Springer, Cham, 2014, pp. 55–72.

[57] J. Song, F. Yang, Y. Zhou, et al., QoE evaluation of multimedia services based on audiovisual quality and user interest, IEEE Trans. Multimedia 18 (3) (2016) 444–457.

[58] M.J. Scott, S.C. Guntuku, Y. Huan, et al., Modelling human factors in perceptual multimedia quality: on the role of personality and culture, in: Proceedings of the 23rd ACM International Conference on Multimedia, 2015, pp. 481–490.

[59] L.R. Goldberg, An alternative" description of personality": the big-five factor structure, J. Pers. Soc. Psychol. 59 (6) (1990) 1216.

[60] E. Goldstein, Sensation and Perception, Cengage Learning, 2013.
[61] G. Hofstede, G.J. Hofstede, M. Minkov, H. Vinken, Values Survey Module 2013, 2013, URL http://www.geerthofstede.nl/vsm2013.
[62] T. Abar, A. Ben Letaifa, S. El Asmi, Real time anomaly detection-based QoE feature selection and ensemble learning for HTTP video services, in: 2019 7th International Conference on ICT & Accessibility (ICTA), IEEE, 2019, pp. 1–6.
[63] V. Menkovski, A. Oredope, A. Liotta, A.C. Sánchez, Predicting quality of experience in multimedia streaming, in: Proceedings of the 7th International Conference on Advances in Mobile Computing and Multimedia, 2009, December, pp. 52–59.
[64] I. Paudel, J. Pokhrel, B. Wehbi, A. Cavalli, B. Jouaber, Estimation of video QoE from MAC parameters in wireless network: A Random Neural Network approach, in: 2014 14th International Symposium on Communications and Information Technologies (ISCIT), IEEE, 2014, September, pp. 51–55.
[65] D.Z. Rodrıguez, R.L. Rosa, G. Bressan, Predicting the quality level of a VoIP communication through intelligent learning techniques, in: The Seventh International Conference on Digital Society (ICDS), 2013, pp. 42–47.
[66] P. Casas, M. Seufert, N. Wehner, A. Schwind, F. Wamser, Enhancing machine learning based QoE prediction by ensemble models, in: 2018 IEEE 38th International Conference on Distributed Computing Systems (ICDCS), IEEE, 2018, July, pp. 1642–1647.
[67] Y.B. Youssef, M. Afif, R. Ksantini, S. Tabbane, A novel QoE model based on boosting support vector regression, in: 2018 IEEE Wireless Communications and Networking Conference (WCNC), IEEE, 2018, April, pp. 1–6.
[68] S. Wassermann, N. Wehner, P. Casas, Machine learning models for YouTube QoE and user engagement prediction in smartphones, ACM SIGMETRICS Perform. Eval. Rev. 46 (3) (2019) 155–158.
[69] J. Jiang, L. Hu, P. Hao, et al., Q-FDBA: improving QoE fairness for video streaming, Multimed. Tools Appl. 77 (9) (2018) 10787–10806.
[70] J.H. Lee, J. Shin, M.J. Realff, Machine learning: overview of the recent progresses and implications for the process systems engineering field, Comput. Chem. Eng. 114 (2018) 111–121.
[71] M. Hall, E. Frank, G. Holmes, et al., The WEKA data mining software: an update, ACM SIGKDD Explor. Newsl. 11 (1) (2009) 10–18.

About the authors

Tasnim Abar is a PhD student at a final stage at Sup'Com Tunisia, she is a teacher at the Higher Institute of Information and Communication Technologies in Tunis. She received her engineer degree in Network and Communication in 2016 from National Engineering School of Gabes. Since 2016, she has been doing research in SDN networks, Quality of Experience QoE, Dynamic adaptive services and ICN networks, she has published five papers in refereed conference proceeding and one journal. Her current research area focuses on D2D communication, IoT, Fog Computing, wireless networks, etc.

Dr. Asma Ben Letaifa is an assistant professor and member of the MEDIATRON research lab at the Higher School of Communications, SUPCOM, University of Carthage, Tunisia. She holds a Telecom Engineering Degree from SUPCOM, University of Carthage, Tunisia and a PhD jointly from SUPCOM, University of Carthage and UBO, Université de Bretagne occidentale, Brest, France. Her research activities focus on telecom services, cloud and mobile cloud architectures, service orchestration, bigData and quality of experience in an SDN/NFV environment. She is author and co-author of several articles on these subjects. She is also author of several courses on telecommunications services, network modeling with queuing theory, web content, cloud architectures and virtualization, massive BigData content and machine learning algorithms. She is also co-author of the "Linux practices" MOOC on the FUN platform.

Prof. Sadok El Asmi is a Professor and member of Cosim research lab at Sup'Com, University of Carthage, Tunisia. He holds his master degree in Applied Mathematics from University of Paris Sud and his PhD in Science from University of Paris Sud, Laboratory of Signals and Systems, CNRS-ESE. His intention is to identify estimation and detection strategies. Targeted applications are biomedical signal processing, QRS complex detection and R-R interval analysis. He has also initiated a research axis combining the algebraic approach with a stochastic approach, called extreme value theory, for the detection of rare events.

Printed in the United States
by Baker & Taylor Publisher Services